Fifty Per Cent
D0189244

Also published in cooperation with the Mathematical Association:

*132 Short Programs for the Mathematics Classroom*

*Maths Talk*

*Sharing Mathematics with Parents — planning school-based events*

*Managing Mathematics — a handbook for the head of department*

# FIFTY PER CENT PROOF

## An Anthology of Mathematical Humour

Compiled by

KEITH SELKIRK
and
WILLIAM WYNNE WILLSON

The Mathematical Association and Stanley Thornes (Publishers) Ltd

First published in 1989 by:
Stanley Thornes (Publishers) Ltd,
Old Station Drive,
Leckhampton,
CHELTENHAM GL53 0DN

British Library Cataloguing in Publication Data

Selkirk, Keith
Fifty per cent proof: an anthology of
mathematical humour
I. Title. II. Willson, William Wynne
828'.91409

ISBN 0 7487 0166 4

Typeset by Tech-Set, Gateshead, Tyne & Wear.
Printed and bound in Great Britain by Ebenezer Baylis & Son Ltd, Worcester.

# PREFACE

Some years ago concern was voiced in the Executive Committee of the Mathematical Association's Teaching Committee about the 'image' of mathematics which is held by the general public. This anthology is intended as a small contribution towards improving this image by giving mathematics a more human face. The idea had two main sources. One was the feeling that many of the snippets which are gathered together in the Association's premier Journal, *The Mathematical Gazette*, and which are called there 'Gleanings', are worthy of a wider audience. The second was the example of that excellent book *A Random Walk in Science*, compiled by R. L. Weber and produced by the Institute of Physics. The Gleanings on their own were insufficiently varied, so we appealed to members of the Association for contributions. Many responded nobly and generously to this request, and we have collected the best of their offerings here together with our own selection of the Gleanings.

The material in this book is humour in its widest sense, intended to amuse rather than to be just 'plain funny'. Most of it is understandable by anyone who has a basic grounding in mathematics, but here and there are pieces which might be incomprehensible to those without sixth form mathematics. We hope these will not deter anyone from enjoying the rest of the book, they are intended as a little bit of spice for those readers whose mathematics is rather more advanced. Some are too good to omit, even at the risk of some mystification. We hope that the material is varied enough to interest everyone.

It is somewhat invidious to name individual contributors, but two are so outstanding that it would be unfair to omit our personal thanks to them for their major contributions. They are A. R. Pargeter who generously contributed a wide personal collection of ideas of an amazingly varied nature, and John Bibby who has most kindly allowed us to use a selection of quotations from his booklet *Quotes, Damned Quotes*, . . . For the cartoons we have drawn particularly from *Manifold*, the much lamented magazine formerly produced by Warwick University. To them, and to all our other contributors, we extend our grateful thanks. Finally we must thank John Hersee for his help and encouragement and the Chairman and Members of the Executive Committee for supporting us in this venture.

One difficulty in compiling an anthology is that of obtaining permission from the copyright holders for the extracts which are reproduced. If there are any errors or omissions in these, or if we have inadvertently omitted to obtain any, then we beg forgiveness and will be happy to make amends in any future edition. Similarly we apologise in advanced for any errors and omissions in attribution, and also to any contributors whose names have not been included.

The compilers have worked together on this book and must take joint responsibility for the results. In practice, Keith Selkirk was reponsible for the text and its selection, while William Wynne Willson was responsible for the cartoons. We hope readers will obtain as much enjoyment from our anthology as we have, and that it will play its part in current efforts to improve the image of mathematics.

Keith Selkirk
William Wynne Willson

#  ACKNOWLEDGEMENTS

The compilers and publishers are grateful to the following for permission to reproduce previously published material:

Andrejs Dunkels
American Scientist
Bryan McAllister
Cambridge University Press
Guardian Newspapers Ltd
IOP Publishing Ltd
Journal of Irreproducible Results
Kluwer Academic Publishers
Macmillan Magazines Ltd
Manifold
Mathematical Association of America
Mathematical Digest, University of Cape Town
Michael Joseph Ltd
The National Council of Teachers of Mathematics
New Scientist
Newspaper Publishing plc
The Observer
The Optical Society of America
Oxford University Press
Private Eye
Punch
Sidney Harris
Times Newspapers Ltd
Unwin Hyman Ltd
The Woburn Press

Every attempt has been made to contact copyright holders, but we apologise if any has been overlooked.

# FIFTY PER CENT PROOF

## You might feel like this about this book . . .

Some years ago there was a Seminar on Control Theory at the University of Leeds. The afternoon session began with a high-powered talk containing some very sophisticated mathematics. After the lecture had been going on for about ten minutes three latecomers entered the room. As luck would have it, the only vacant seats were in the middle of the centre row, and a few minutes passed before the latecomers were able to squeeze their way to these empty seats.

Everyone settled down again and the high-powered lecture continued. It was noticed that the three latecomers seemed not at ease, and they began whispering to each other. After a few more minutes one of them stood up, interrupted the lecture and asked "Excuse me, but are we at the right lecture?" To everyone's embarrassment the answer was "Yes", they were at the right lecture.

**Alan Slomson, University of Leeds.**

You are reading the right book,
so settle comfortably in your seat and enjoy it!

## Bertrand Russell's dream

I can remember Bertrand Russell telling me of a horrible dream. He was on the top floor of the University Library, about A.D. 2100. A library assistant was going round the shelves carrying an enormous bucket, taking down book after book, glancing at them, restoring them to the shelves or dumping them into the bucket. At last he came to three large volumes which Russell could recognise as the last surviving copy of *Principia Mathematica*. He took down one of the volumes, turned over a few pages, seemed puzzled for a moment by the curious symbolism, closed the volume, balanced it in his hand and hesitated . . .

**G. H. Hardy, *A Mathematician's Apology*, 1948,**
**reproduced by kind permission of the Cambridge University Press.**

## Hot . . .

A young man approached us. He was wrapped up against the biting 38 degrees centigrade cold, his face hardly visible.

**Spotted by E. K. Lloyd in *The Southern Evening Echo*, 23 January 1987.**

## . . . and cold

In Leningrad, in December, beset by wet snow and Baltic gales, the number of hours of sunshine per day averages an appalling zero.

**Seen in a guidebook to Russia, Graham Jones.**
***MG*, 1987, pp. 134–5.**

## Just a second

A circle is divided into 360 degrees, each degree into 60 minutes, and each minute into 60 seconds; so one second is 1/1000 of a circle.

**Michie and Johnston, *The Creative Computer*.**
**[Per John Owen.] *MG**, 1986, p. 271.**

## Overkill

Post Office staff began to deliver 23m government AIDS campaign leaflets to every UK household.

***The Financial Times*, 13 January 1987, sent in by Simon Gray.**

Statistical projections from such figures have been used to indicate that by the year 2000 the number of AIDS cases could be 100 million times the population of the world, but among official bodies only West Germany's Ministry of Health believed this possible.

***The Times*, 6 March 1987. [Per David Blackman.] *MG*, 1987, p. 126.**

*Throughout the book, the abbreviation *MG* refers to *The Mathematical Gazette*.

## The Major-General's song

I am the very model of a modern Major-General,
I've information vegetable, animal and mineral,
I know the kings of England and I quote the fights historical,
From Marathon to Waterloo in order categorical;
I'm very well acquainted too with matters mathematical,
I understand equations, both the simple and quadratical,
About binomial theorem I'm teeming with a lot of news —
With many cheerful facts about the square on the hypotenuse.
I'm very good at integral and differential calculus,
I know the scientific names of beings animalculous;
In short, in matters vegetable, animal and mineral,
I am the very model of a modern Major-General.

**W. S. Gilbert, *The Pirates of Penzance*. Contributed by Frank Chorlton, University of Aston, and others.**

## Design fault

The rotors [of the Chinook helicopter] are designed to intersect.

**Seen in *The Observer*, 9 November 1986 by R. S. J. Good. *MG*, 1987, p. 292**

## A boxing match

"64 identical wooden cubes fit exactly into a transparent cubical box. How many of them are visible?" I once set this as a school entrance scholarship question. The correct answer is, of course 56. I had to allow 52 in the case of one candidate who was too lazy to pick it up off the table on which he imagined it to be lying. I was delighted with two candidates who gave the answer as 96. Explain!

**Contributed by A. R. Pargeter, Devon.**

## Jam side down again!

[Henry Kissinger relates what happened when he and President Nixon boarded a Soviet plane in Kiev and the engines refused to start.]

While a backup plane was being readied Kosygin stormed onto the plane and said: "Tell us what you want with our Minister of Aviation. If you want him shot on the tarmac we will do so."

He looked as if he might be serious. I attempted to ease his embarrassment by speaking of the wickedness of objects. If one dropped a piece of toast it would fall on the buttered side in direct proportion to the value of the rug, I said; when one dropped a coin it always rolled away, never towards one. Kosygin was not to be consoled by such transparent attempts to shift responsibility. "This is not my experience," he said, fixing me with a baleful glance. "I have dropped coins which rolled towards me."

**From Henry Kissinger, *The White House Years*, 1979, quoted in *The Sunday Times*, 16 December 1979,** 
**Contributed by E. Gulbenkian, Surrey.**

## Two surprising results

If ABC is a triangle, $r$ is the radius of its incircle and $r_1$, $r_2$ and $r_3$ are the radii of its three excircles, then:

(1) The triangle is right angled if and only if

$$r + r_1 + r_2 + r_3 = a + b + c$$

(2) The triangle is right angled at C if and only if

$$r_1 r_2 = r_3 r$$

**Contributed by A. R. Pargeter.**

## How odd!

At a party, the number of people who shake hands an odd number of times is even.

**Contributed by A. R. Pargeter.**

## Are they really solutions?

Solve $\dfrac{2x+1}{(x+3)(x-2)} + \dfrac{5x}{(2x+1)(x+3)} = \dfrac{x+3}{(2x+1)(x-2)}$.

[There are no solutions.]

**Contributed by A. R. Pargeter.**

## Almost congruent

How is it possible for two triangles to be identical in five respects (i.e. sides and angles), and yet not to be congruent?

[Take the sides to be $a^3$, $a^2b$ and $ab^2$ for the first triangle and $a^2b$, $ab^2$ and $b^3$ for the second triangle.]

**Contributed by A. R. Pargeter.**

***Manifold*, Spring 1975.**

## Below one

I once mentioned to a class that I knew of a road with a house numbered 0. We discussed the (pretty obvious) reason for this, and I asked "If another house were built at the beginning of the road, how should we number it?" Whereupon, as quick as a flash, one pupil replied "Sir, if it was a public house it would be Bar One!"

What I call pips, my wife calls poppers, and the manufacturers call snap fasteners come in sizes 5, 4, 3, 2, 1, 0, 00, 000, . . . — the range not having been properly anticipated at the outset! But the lesson has not been learnt — padded ("Jiffy") envelopes also go down to size 000!

**Contributed by A. R. Pargeter.**

## Sounds logical . . .

The law of the excluded middle either rules or does not rule, O.K.?

**Contributed by John Bibby, *Quotes, Damned Quotes, . . .***

## The mathematicians' Christmas greeting

$$\frac{(\pi/2 - C_1)(\pi/2 - C_2)(\pi/2 - C_3)\ldots(\pi/2 - C_n)}{2u}$$

[The complements of the $C$'s on two $u$.]

**Contributed by G. Tyson, London.**

## Union of two fractions?

*Giuseppe:* It's quite simple. Observe. Two husbands have managed to acquire three wives. Three wives — two husbands. That's two thirds of a husband to each wife.

*Tessa:* O Mount Vesuvius, here we are in arithmetic! My good sir, one can't marry a vulgar fraction!

*Giuseppe:* You've no right to call me a vulgar fraction.

**W. S. Gilbert, *The Gondoliers*. Contributed by Frank Chorlton.**

## The chain rule

It is easy to form a chain from a string of zeros.

**Stanislaw Jerzy Lec, *Unkempt Thoughts*, St Martin's Press, 1962, quoted in Clifton Fadiman, *The Mathematical Magpie*, Simon and Schuster, New York, 1962.**

# Quickest way to a million

**People who do very unusual jobs indeed (30): The man who counts people at public gatherings, and everything else as well.**

You've probably seen his headlines, "Two million flock to see Pope." "200 Arrested as Police Find Ounce of Cannabis." "Britain £3 billion in red." You probably wondered who was responsible for producing such well rounded-up figures. What you didn't know was that it was all the work of one man, Rounder-Up to the Media, John Wheeler. But how is he able to go on turning out such spot-on statistics? How can he be so accurate all the time?

"We can't" admits Wheeler blithely. "Frankly, after the first million we stop counting, and we round it up to the next million, I don't know if you've ever counted a papal flock, but not only do they all look a bit the same, they also don't keep very still, what with all the bowing and crossing themselves.

"The only way you could do it accurately is taking an aerial photograph of the crowd and hand it to the computer to work out. But then you'd get a headline saying "1,678,1634 *[sic]* flock to see Pope, not including 35,467 who couldn't get a glimpse of him", and, believe me, nobody wants that sort of headline."

The art of big figures, avers Wheeler, lies in psychology, not statistics. The public likes a figure it can admire. It likes millionaires, and million-sellers, and centuries at cricket, so Wheeler's international agency gives them the figures it wants, which involves not only rounding up but rounding *down.*

"In the old days people used to deal with crowds on the Isle of Wight principle. You know, they'd say that every day the population of the world increased by the number of people who could stand upright on the Isle of Wight, or the rain forests were being decreased by an area the size of Rutland. This meant nothing. Most people had never been to the Isle of Wight for a start, and even if they had, they only had a vision of lots of Chinese standing in the grounds of the Cowes Yacht Club. And the Rutland comparison was so useless that they were driven to abolish Rutland to get rid of it.

"No, what people want is a few good millions. A hundred million, if possible. One of our inventions was street value, for instance. In the old days they used to say that police had discovered drugs in a quantity large enough to get Rutland stoned for a fortnight. *We* started saying that the drugs had a street value of £10 million. Absolutely meaningless, but people understand it better."

Sometimes they do get figures spot on. "250,000 flock to see royal two", was one of his recent headlines, and although the 250,000 was a rounded up figure, the two was quite correct. In his palatial office he sits surrounded by relics of past headlines — a million-year-old fossil, a £500,000 Manet, a photograph of Mrs Thatcher's £500,000 house — but pride of place goes to a pair of shoes framed on the wall.

"Why the shoes? Because they cost me £39.99. They serve as a reminder of mankind's other great urge, to have stupid odd figures. Strange, isn't it? They want mass demos of exactly half a million, but they also want their gramophone records to go round at $33\frac{1}{3}$, 45 or 78 rpm. We have stayed in business by remembering that below a certain level people want oddity. They don't want a rocket costing £299 million and 99p, and they don't want a radio costing exactly £50."

How does he explain the times when figures clash – when, for example, the organisers of a demo claim 250,000 but the police put it at nearer 100,000?

"We provide both sets of figures, the figures the organisers want and those the police want. The public believes both. If we gave the true figure, about 167,890, nobody would believe it because it doesn't really sound believable."

John Wheeler's name has never become well known, as he is a shy figure, but his firm has an annual turnover of £3 million and his eye for the right figure has made him a very rich man. His chief satisfaction, though, comes from the people he meets in the counting game.

"Exactly two billion, to be precise."

**Miles Kington in *The Observer*, 3 November 1986, contributed by Brian Cooper, and reproduced by kind permission of *The Observer*.**

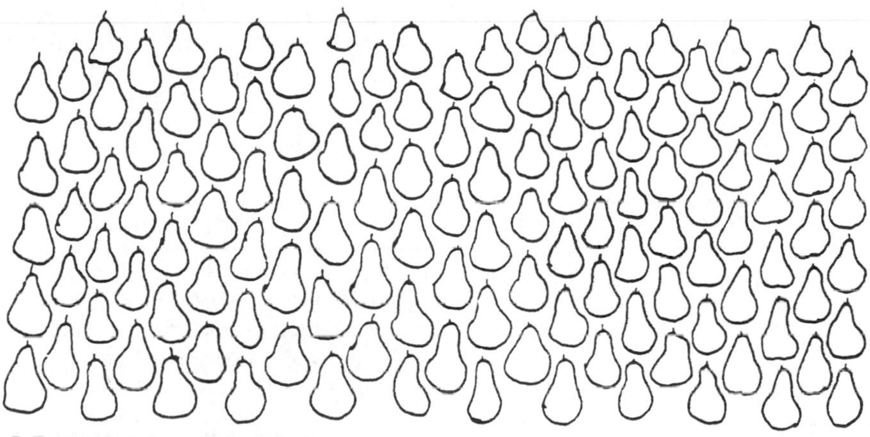

ORDERED PEARS

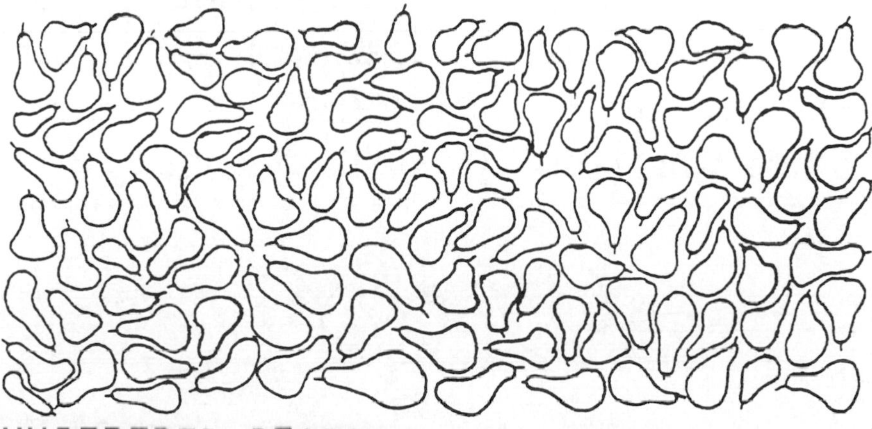

UNORDERED PEARS

***Manifold*, Spring 1980.**

## The ultimate principle

This principle is so perfectly general that no particular application of it is possible.

G. Polya, *How to Solve it*, Princeton University Press, 1945, quoted in Clifton Fadiman, *The Mathematical Magpie*, Simon and Schuster, New York, 1962.

## Fractional drunkenness

Four cases of drunkenness have been reported during the past year: not serious you would say, but it represents a 50 per cent increase.

'Country Sessions', in *The Countryman*, Summer 1955. [Per A. R. Pargeter.] *MG*, 1956, p. 180.

## Base five

I went this far with him: "Sir, allow me to ask you one question. If the church should say to you, 'Two and three make ten,' what would you do?" "Sir," said he, "I should believe it, and I should count like this: one, two, three, four, *ten*." I was now fully satisfied.

Boswell's Journal for 31 May, 1764 in *Boswell in Holland*, ed. F. A. Pottle, Yale University Press, 1952, p. 258. [Per Professor E. H. Neville.] *MG*, 1956, p. 14.

## Roots and branches

The different branches of Arithmetic — Ambition, Distraction, Uglification, and Derision.

Lewis Carroll, *Alice in Wonderland*.

## Motion in a circle

The Nullabor plain on the east–west line from Port Augusta to Kalgoorlie 297.3 miles of [railway] line absolutely straight except that the earth's curvature is said to make it twelve feet higher at one end than the other.

*The Listener*, 26 March 1953, p. 515. [Per A. R. Pargeter.] *MG*, 1957, p. 139.

Andrejs Dunkels, contributed by Stig Olsson, Sweden.

## Not-so-random walks

Contending that drunken people automatically stagger eastwards, the Maryland police publication recommended that patrol cars should park to the east of any "drunk" about to be arrested. It cited as its authority a Dr. C. Volk, of California, who attributed this peculiarity to the spin of the earth.

He said babies learning to walk also toddled towards the east. Washington police say if anything drunks stagger to the west since the right leg is usually the stronger.

*Daily Telegraph*, 30 December 1954. [Per E. A. Side.] *MG*, 1955, p. 279.

## False premises?

Geometry is the art of correct reasoning on incorrect figures.

G. Polya, *How to Solve It*, 1945, Princeton University Press, quoted in Clifton Fadiman, *The Mathematical Magpie*, Simon and Schuster, New York, 1962.

## We know the feeling!

I wish in conclusion to express my humble thanks to Professor Littlewood, but for whose patient profanity this paper could never have become fit for publication.

S. Skewes, *Proc. London Math. Soc.* (3) V, March 1955, p. 50. [Per Professor E. H. Neville.] *MG*, 1956, p. 42.

## Whatever happened to the rod, pole or perch?

It has been recommended that Britain change to the Metric System. What nonsense is this? Britain and the Commonwealth were built by enterprise and industry founded on the bedrock of British weights and measures.

Our job is to *lead* the Continent — not follow it. The Metric System did not save France in 1940!

*Daily Mail* [Per R. F. Wheeler.] *MG*, 1956, p. 259.

## A trigonometric result

$$\frac{\sin^4\theta + \cos^4\theta}{4} \text{ and } \frac{\sin^6\theta + \cos^6\theta}{6}$$

differ by the constant amount $\frac{1}{12}$.

[Note: integrate $\sin^3\theta\cos^3\theta$.]

Contributed by A. R. Pargeter.

## Where's the proof?

A mathematician had a horseshoe over his office door. A colleague said that he thought the mathematician did not believe in such things. The mathematician replied that that was true, but he had been told that it would still work even if he did not believe in it.

Contributed by Keith Austin.

# Heaven is hotter than Hell

The temperature of Heaven can be rather accurately computed from available data. Our authority is the Bible: Isaiah 30:26 reads, *Moreover the light of the Moon shall be as the light of the Sun and the light of the Sun shall be sevenfold, as the light of seven days*. Thus heaven receives from the Moon as much radiation as we do from the Sun and in addition seven times seven (forty-nine) times as much as the Earth does from the Sun, or fifty times in all. The light we receive from the Moon is a ten-thousandth of the light we receive from the Sun, so we can ignore that. With these data we can compute the temperature of Heaven. The radiation falling on Heaven will heat it to the point where the heat lost by radiation is just equal to the heat received by radiation. In other words, Heaven loses fifty times as much heat as the Earth by radiation. Using the Stefan–Boltzmann fourth-power law for radiation

$$(H/E)^4 = 50$$

where $E$ is the absolute temperature of the Earth, 300 K. This gives $H$ as 798 K (525 °C).

The exact temperature of Hell cannot be computed but it must be less than 444.6 °C, the temperature at which brimstone or sulphur changes from a liquid to a gas. Revelation 21:8: *But the fearful, and unbelieving . . . shall have their part in the lake which burneth with fire and brimstone.* A lake of molten brimstone means that its temperature must be below the boiling point, which is 444.6 °C. (Above this point it would be a vapour, not a lake.)

We have, then, temperature of Heaven, 525 °C. Temperature of Hell, less than 444.6 °C. Therefore, Heaven is hotter than Hell.

***Applied Optics*, 11, A14, (August 1972), quoted in *A Random Walk in Science*, ed. R. L. Weber, Institute of Physics, London, 1973.**

**Sidney Harris, contributed by Stig Olsson.**

## Modern research in mathematics

The advent of Modern Mathematics in the educational limelight has produced an interest in the work of the professional mathematician and in the question, "How can one do research in mathematics?" The following passage formed the introduction to a recent paper and indicates to some extent the line taken by a number of mathematicians.

### A note on Piffles by A. B. Smith

A. C. Jones in his paper "A Note on the Theory of Boffles", Proceedings of the National Society, 13, first defined a Biffle to be a non-definite Boffle and asked if every Biffle was reducible.

C. D. Brown in "On a paper by A. C. Jones", Biffle, 24, answered in part this question by defining a Wuffle to be a reducible Biffle and he was then able to show that all Wuffles were reducible.

H. Green, P. Smith and D. Jones in their review of Brown's paper, Wuffle Review, 48, suggested the name Woffle for any Wuffle other than the non-trivial Wuffle and conjectured that the total number of Woffles would be at least as great as the number so far known to exist. They asked if this conjecture was the strongest possible.

T. Brown in "A Collection of 250 Papers on Woffle Theory dedicated to R. S. Green on his 23rd Birthday" defined a Piffle to be an infinite multi-variable sub-polynormal Woffle which does not satisfy the lower regular Q-property. He stated, but was unable to prove, that there were at least a finite number of Piffles.

T. Smith, L. Jones, R. Brown and A. Green in their collected works "A short introduction to the classical theory of the Piffle", Piffle Press, 6 gns., showed that all bi-universal Piffles were strictly descending and conjectured that to prove a stronger result would be harder.

It is this conjecture which motivated the present paper.

**By Keith Austin, University of Sheffield, in *MG*, 1967, pp. 149–50.**

## Not if you put the lid on . . .

A class were trying to find the volume of water in a swimming pool whose bottom was not a single plane. The teacher had to tell the pupils to use $V = Ah$ for a prism, taking the side of the pool as the base. One pupil asked "Wouldn't all the water fall out?"

**Contributed by John MacNeill.**

## A working approximation

The first law of Engineering Mathematics: All infinite series converge, and moreover converge to the first term.

**Anon., contributed by Martin J. Savier (?), University College, Aberystwyth.**

*New Scientist.*

## Quotelets from Bibby (*Quotes, Damned Quotes, . . .*)

If there is a 50–50 chance that something will go wrong, then 9 times out of 10 it will.

*Paul Harvey News*, Fall 1979.

Counting in octal is just like counting in decimal if you don't use your thumbs.

Tom Lehrer.

It is now proved beyond doubt that smoking is one of the leading causes of statistics.
(This law is known as Knebel's Law.)

John Peers, *1001 Logical Laws*, Hamlyn, 1979.

Tell me not, in mournful numbers, Life is but an empty dream.

H. W. Longfellow.

I am one of the unpraised, unrewarded millions without whom statistics would be a bankrupt science. It is we who are born, who marry, who die, in constant ratios.

Logan Pearsall Smith.

We must believe in luck. For how else can we explain the success of those we don't like?

Jean Cocteau.

The figure of 2.2 children per adult female was felt to be in some respects absurd, and a Royal Commission suggested that the middle classes be paid money to increase the average to a rounder and more convenient number.

*Punch*, quoted in M. J. Moroney, *Facts from Figures*, Penguin, 1951.

# More quotelets from Bibby (*Quotes, Damned Quotes, . . .*)

Life is the art of drawing sufficient conclusions from insufficient premises.

**Samuel Butler.**

I find a coin in the street, flip it twice, and both times it comes up heads. The maximum likelihood estimate is that this coin has two heads. The Bayesian estimate is that it has one head. In this case, which is better?

**R. J. Wonnacott and T. W. Wonnacott, *Introductory Statistics*, Wiley, second edition, 1972.**

Nothing is so fallacious as fact, except figures.

**George Canning.**

Figures won't lie, but liars will figure.

**General Charles H. Grosvenor.**

He uses statistics as a drunk uses a street lamp, for support rather than illumination.

**Andrew Lang.**

## Who was Cicero?

James Killian expressed it (the increasing separateness of science in the modern world) more bluntly by reporting two intelligences that have been making the rounds of the faculty lounges — one the observation that the scientist knows nothing of the liberal arts and regrets it, while the humanist knows nothing of science and is proud of it. The other was an incident said to have occurred at a liberal arts faculty meeting. When a student named Cicero was reported to have flunked Latin, everybody laughed, but when a student named Gauss was reported to have failed mathematics, only the science professors laughed.

**William S. Beck, *Modern Science and the Nature of Life*. [Per Dr H. Martyn Cundy.] *MG*, 1959, p. 109.**

## A recurring transcendental

About the only factual question I remember in the whole series was one asking the mathematical meaning of 'pi'. Gould explained that it represented the ratio between the radius and the circumference of a circle. Asked by McCullough if he could tell listeners what exactly the figure was he said that it roughly could be represented by the fraction twenty-two over seven. He went on to say that as far as it could be calculated it was three point . . . and he then added thirty-five more figures. McCullough took these down on a piece of paper as he spoke them and at the finish said, 'Gould, can you repeat those figures?' 'Certainly,' said Gould, and proceeded to do so. I was not so impressed afterwards, however, when I worked it out myself. 'Pi' is represented by 3.142857 with the last six figures repeated to infinity!

**Commander A. B. Campbell, R. D., *When I was In Patagonia*, 1953. [Per N. Cannell and G. A. Garreau.] *MG*, 1957, p. 253.**

# Pas devant les enfants

It's not television which is the greatest threat to the art of conversation, in my experience; it's children.

The Victorians were certainly 100 years ahead of their time on this problem, with their doctrine of children being seen and not heard. If the children are heard, then any adults present are not; and the Victorians, with their usual sensitive concern for parent-care, realised that the most frightful psychic damage could be inflicted upon adults whose natural drive to communication and self-expression was persistently frustrated.

What I don't quite understand, though, is the Victorians' inexplicable permissiveness in letting children be *seen*. A child, at any rate a small child, doesn't need to be heard to disrupt all rational intercourse in the vicinity; its visible presence is quite enough.

A child is rather like a television set, in fact — and turning the sound down when company comes isn't enough to prevent all eyes in the room from being irresistably drawn towards it. How often has one seen a whole room full of normally articulate adults sitting bemused around their assembled offspring, bereft of all powers of speech, apart from the sort of desultory facetious comment which is usually reserved for the Westerns? And television sets have the great advantage that they can be switched off.

Of course, if children are disruptive with the sound turned down, they're worse still when it's turned up. Sound and vision tend to be deployed to their fullest communication-destroying effect just when a matter of some delicacy and importance has to be conveyed between husband and wife. Just when Mr Ricardo was trying to explain to Mrs Ricardo about Marginal Utility — that's when the children would have switched on the jammers; just when the Tsar and Tsarina were discussing whether to invade Poland next season.

Or take the afternoon that Pythagoras came out of his study, looking rather pleased with himself.

'You know this work I've been doing recently on hypotenuses?' he says to his wife, trying to sound casual. 'Well, a rather interesting point struck me this afternoon — I don't know whether you think it sounds reasonable — that the *square* on the hypotenuse must be equal to . . .'

But at this point a ringing cry from the lavatory interrupts the exposition.

'Will you wipe my b-o-o-o-t-tom!'

'Sorry,' says his wife when she gets back. 'What were you saying? Something about hypotenuses.'

'I was just going to say that the square on the hypotenuse . . .'

'Mummy!'

'Sh, Jemima! Daddy's talking.'

'. . . that the square on the hypotenuse is equal to . . .'

'But, *Mummy* . . .!'

'Sh, Jemima! You mustn't interrupt when someone's speaking! How many times have I had to tell you?'

'. . . equal to the sum of the squares on the other two sides!"

'I see. Now what's the trouble, Jemima?'

'James is being horrible to me! He's taken my zoetrope!'

'James, give Jemima back her zoetrope at once! Sorry, Py. What were you saying?'

'I said it.'

'All hypotenuses are equal . . .?'

'God give me strength! Why do you *never* listen to what I say? I said the square on the hypotenuse is equal to the sum . . .'

'The what?'

'The *sum* . . . the SUM . . .! My God, I can't hear myself speak! Will you SHUT UP you two! If I hear one more word out of either of you, I'll throw that damned zoetrope into the Aegean, and that'll be the end of it! Now, the *square* on the *hypotenuse* . . .'

'Yes, yes, I got that bit . . . Just a moment — James, what *have* you been doing to your face . . .? Well, go and wash it off at once . . . Sorry — "the square on the hypotenuse" — I am listening . . . Don't just rub it on your sleeve, James . . .! Sorry, Py, but if he's left to go wandering around in that condition there'll be shaving cream all over the house . . . Anyway the square on the hypotenuse . . .'

'Will you wipe my b-o-o-o-o-t-tom!'

It's Pythagoras' turn this time. 'Where are we?' he asks wearily when he returns. 'Oh, yes, the square on the hypotenuse. Well, all I was going to say was that it's equal to the sum of the squares . . . *Now* what are you doing? What the hell do you keep turning round for?

'Sorry, — I was just trying to see why Jemima was so quiet all of a sudden.'

'Oh, for God's sake!'

'Go on about the square on the hypotenuse.'

'It was nothing.'

'Don't be silly.'

'It wasn't of the slightest importance . . . Well, I was merely going to say that it was equal to the sum of the squares on the other two sides. That's all.'

'But, Py, that's absolutely *fascinating!* I'd never have guessed it! Marvellous . . .! What *is* Jemima up to by the way? Is she sulking? Can you see? She's not sucking her thumb is she?'

'Yes . . . No . . . I don't know! She's not there . . . Look, are you interested in my work on the hypotenuse or aren't you?'

'Of course I am. I think it's tremendously important . . . I'd better just make sure she hasn't wandered out into the street . . .'

'I mean, I don't care whether you are or not. I just thought you *were*, that's all. It's just that once upon a time you used to ask me . . .'

'Will you wipe my b-o-o-o-o-t-tom!'

It's the trailing clouds of glory which Wordsworth observed hanging about children — that's what really disrupts communication. Glorious-looking clouds, certainly; but when you're in amongst them, like most clouds, pretty well indistinguishable from dense fog.

**Michael Frayn in *The Observer*, 1968, contributed by Neil Bibby, King's College, London, and reproduced by permission of *The Observer*.**

## Independence, or, is your axiom really necessary?

**Candidates offering History at A level must take Paper I, and two of the Papers II, III, IV or V, with the following restrictions:**

**If IV is taken the third paper must be II.**

**If V is taken the third paper must be either II or III.**

**From the *Regulations* of the Oxford and Cambridge Schools Examination Board. [Per R. F. Wheeler.] *MG*, 1960, p. 34.**

## Numbest?

**Numb**, adj., devoid of sensation . . .
**Number**, comparative of numb.

**Webster's *Third New International Dictionary*. Contributed by Keith Selkirk.**

## "If" (school certificate maths version)

If you can solve a literal equation
And rationalise denominator surds,
Do grouping factors (with a transformation)
And state the factor theorem in words;
If you can plot the graph of any function
And do a long division sum (with gaps),
Or square binomials without compunction,
Or work cube roots with logs without mishaps.
If you possess a sound and clear-cut notion
Of interest sums with *P* and *I* unknown;
If you can find the speed of trains in motion,
Given some lengths and "passing times" alone;
If you can play with *R* (both big and little)
And feel at home with *l* (or *h*) and $\pi$,
And learn by cancellation how to whittle
Your fractions down till they delight the eye.
If you can recognise the segment angles
Both at the centre and circumference;
If you can spot equivalent triangles
And Friend Pythagoras (his power's immense);
If you can see that equiangularity
And congruence are two things and not one,
You may pick up a mark or two in charity,
And, what is more, you may squeeze through, my son.

**IVH, in *The Times Educational Supplement*, 19 July 1947, contributed by A. R. Pargeter, and reproduced by permission of the *TES*.**

## A meal for Emile

I shall never forget seeing a young man at Turin, who had learnt as a child the relations of contours and surfaces by having to choose every day isoperimetric cakes among cakes of every geometrical figure. The greedy little fellow had exhausted the art of Archimedes to find out which were the biggest.

**J. J. Rousseau, *Emile*, p. 111. [Per L. W. H. Hull.] *MG*, 1958, p. 108.**

## A powerful conclusion

A lecturer at King's College, London, who claimed not to have had the benefit of a classical upbringing had always left off 'Q.E.D.' as a somewhat mysterious phrase and one showing an unwelcome degree of snobbishness. He preferred to end his proofs with $W^5$, which was an abbreviation for 'Which was what was wanted'.

**Contributed by Matthew Linton, University of Hong Kong.**

## Victorian problems

(1) A gentleman going into a garden, meets with some ladies, and says to them, Good morning to you 10 fair maids. Sir, said one of them, we are not 10; but if we were twice as many more as we are, we should be as many above 10 as we are now under — how many were they?
[*Answer:* 5.]

(2) What length of cord will be fair to tie to a cow's tail, the other end fixed in the ground, to let her have the liberty of eating an acre of grass and no more, supposing the cow and tail to be $5\frac{1}{2}$ yards?
[*Answer:* 6.136 perches.]

**Promiscuous (i.e. miscellaneous) questions**

(3) How long would a railway engine be in going to the sun, at the rate of 40 miles per hour, reckoning the sun's distance 95,000,000 miles?
[*Answer:* 271 years, 43 days, 8 hours.]

(4) From £100 borrow'd, take £70 paid. 'Twas a virgin that lent it — what's due to the maid?

(5) A sheepfold was robbed three nights successfully; the first night half the sheep were stolen, and half a sheep more; the second night half the remainder were lost, and half a sheep more; the last night they took half that were left, and half a sheep more; by which time they were reduced to 20 — how many were there at first?

(6) A person had 20 children, and there was $1\frac{1}{2}$ years between each of their ages; his eldest son was born when he was 24 years old, and the age of the youngest is 21 — what was the father's age?

(7) If the tap of a large cistern will empty it in 37 minutes, how many of such taps will empty it in $6\frac{1}{2}$ minutes?
[*Answer:* five and nine-thirteenths.]

**Some problems from Nicholson's *Walkingame's Arithmetic, simplified according to modern advancement in arithmetical science*, 1852 edition. Contributed by G. Thompson, Peterborough.**

## Newton and Einstein

Nature and Nature's Laws lay hid from sight
God said *Let Newton be!* and all was light.

**Alexander Pope, contributed by D. Brown, York.**

It did not last: the Devil howling 'Ho!
Let Einstein be!' restored the status quo.

**Sir John Collings Squire, contributed by Keith Selkirk.**

## Permissive society

In the thirties, Einstein said in England you were allowed to do anything you were not forbidden to do; in Germany, you were forbidden to do anything you were not allowed to do; in Austria, you were allowed to do anything you were forbidden to do.

**From a letter to *The Independent*, 19 March 1988, by J. R. Hawthorn, Herefordshire, contributed by Keith Selkirk, and reproduced by permission of *The Independent*.**

***Manifold*, Autumn 1975.**

## Quotelets from Bibby (*Quotes, Damned Quotes, . . .*)

All research can be classified under one of three headings: "the proof of the blindingly obvious", i.e. proving what was already known; "the great leap sideways", i.e. towards an irrelevant or unjustified conclusion, not part of the original hypothesis; the "we'll prove it if it kills you" attitude, i.e. presenting a mass of incomprehensible statistics intended to overcome any criticism by quantity alone.

*The Times* (via B. G. Birdseye)

What used to be called prejudice is now called a null hypothesis.

A. W. F. Edwards, *Nature*, 9 March 1971.

Figures often beguile me, particularly when I have the arranging of them myself; in which case the remark attributed to Disraeli would often apply with justice and force: "There are three kinds of lies: lies, damned lies and statistics".

Mark Twain, *Autobiography*, (p. 149 in the 1960 edition).

A statistician is a person who draws a mathematically precise line from an unwarranted assumption to a foregone conclusion.

Anon.

Public ignorance of the laws of evidence and of statistics can hardly be exaggerated.

G. B. Shaw.

# Busprongs

It will be assumed that the reader is familiar with the terms and results of the elementary theory of busprongs. Whenever a result for busprongs depends on a result from the elementary theory of prongs, then the details of the logical deduction will be left to the reader.

DEFINITION. Let $P$ be a prong and let $b$ be a clackle of $p$. We say $b$ is a *busprong* of $p$ if the following conditions are satisfied:

(BP 1) (Hafulness) $b$ is haful.

(BP 2) (Shutter's Criterion)
$\overline{w_1; B; w_2}$ for all $w_1$, $w_2$.

(BP 3) (Podanti's Condition)
$\binom{B}{\lambda}$ for all $\lambda$

We are immediately faced with the question of whether busprongs exist and, if they do, whether they are plentiful or scarce. The following result goes some way to answering these questions and indicates the richness of the theory of busprongs.

THEOREM 1. Let $P$ be a prong.

(a) $\langle P \rangle$ is a busprong.

(b) $P_P^P$ is a busprong.

*Proof*

(a) (BP 1) $\langle P \rangle$ is haful by the definition of prong.

(BP 2) Let $w_1$, $w_2$ be given. Then $\overline{w_1; \langle P \rangle; w_2}$ follows from the definition of prong.

(BP 3) Let $\lambda$ be given. Then $\binom{\langle P \rangle}{\lambda}$ follows from the definition of prong.

Thus $\langle P \rangle$ is a busprong.

(b) (BP 1) $P_P^P$ is haful by the definition of $P_P^P$.

(BP 2) Let $w_1$, $w_2$ be given. Then $\overline{w_1; P_P^P; w_2}$ follows from the definition of $P_P^P$.

(BP 3) Let $\lambda$ be given. Then $\binom{P_P^P}{\lambda}$ follows from the definition of $P_P^P$.

Thus $P_P^P$ is a busprong. $\triangle$

*Note.* Although we have actually proved $\langle P \rangle$ is a busprong we will usually say that $P$ is a busprong, provided there is no danger of confusion. This convention brings us into line with the common usage of saying that $\langle P \rangle$ is a prong when we actually mean that $P$ is a prong. The next theorem shows the well-behavedness of busprongs.

THEOREM 2. Let $B_1$, $B_2$ be busprongs. Then $(B_1 \rightarrow B_2)$ is a busprong.

*Proof*
(BP 1) $(B_1 \rightarrow B_2)$ is haful follows from the fact that $B_1$ and $B_2$ are haful by (BP 1) and the definition of $(B_1 \rightarrow B_2)$.
(BP 2) Let $w_1$, $w_2$ be given. Then $\overline{w_1; B_1; w_2}$ and $\overline{w_1; B_2; w_2}$ by (BP 2) and so by the definition of $(B_1 \rightarrow B_2)$ it follows that $\overline{w_1; (B_1 \rightarrow B_2); w_2}$.
(BP 3) Let $\lambda$ be given. Then $\begin{pmatrix} B_1 \\ \lambda \end{pmatrix}$ and $\begin{pmatrix} B_2 \\ \lambda \end{pmatrix}$ by (BP 3) and so by the definition of $(B_1 \rightarrow B_2)$ it follows that $\begin{pmatrix} (B_1 \rightarrow B_2) \\ \lambda \end{pmatrix}$.
Thus $(B_1 \rightarrow B_2)$ is a busprong. $\triangle$

DEFINITION. Let $B$ be a busprong and $K$ be a kormile. The *frigate* of $B$ and $K$, $\{B - K\}$ is defined to be $[[B, K]]$. Similarly the *grifate* of $K$ and $B$, $\{K - B\}$ is defined to be $[[K, B]]$.

DEFINITION. A busprong $B$ is said to be a *nanu-busprong* if

$$\begin{bmatrix} \{B - K\} \\ \{K - B\} \end{bmatrix}, \quad \text{for all } K.$$

DEFINITION. Let $P$ be a prong and $B$ be a nanu-busprong. Let

$$\alpha = (\{B - K\} \rightarrow K \rightarrow P) \text{ and}$$
$$\beta = (\{B - K_1\}; \{B - K_2\} \rightarrow \{B - K_1; K_2\} \rightarrow P).$$

Then $[\alpha - \beta]$ is called a *quiprong* and is denoted by $P \downarrow B$.
*Note.* It is necessary to show that $P \downarrow B$ is well-defined and that we do in the first part of the next theorem.

THEOREM 3. Let $P$ be a prong and $B$ be a nanu-busprong.

(a) Let $K_1$, $K_2$, $K_3$, $K_4$ be kormiles such that

$$\begin{bmatrix} \{B - K_1\} \\ \{B - K_2\} \end{bmatrix} \text{ and } \begin{bmatrix} \{B - K_3\} \\ \{B - K_4\} \end{bmatrix} \text{ then } \begin{bmatrix} \{B - K_1; K_3\} \\ \{B - K_2; K_4\} \end{bmatrix}.$$

(b) $P \downarrow B$ is a prong.

The proof is left to the reader. As a hint the reader is reminded not only of the nanu-ness of $B$ but also the busprong-ness of $B$. Also for (b) it will be helpful to have the definition of a prong to hand.

My first reaction to the above is that any student presented with this as a course in pure mathematics is not learning about busprongs but is performing logical operations. Perhaps I am being too harsh and my second reaction is that the student is being presented with a list of facts about busprongs together with a chain of logical steps connecting the facts. It is like the old style geography lesson when the teacher listed the towns and rivers of a country, logically arranged in descending size, but did not show even so much as a photograph of the country to indicate what it was like. It is as if we replace a visit to London by a reading of its telephone directory, which is so logically arranged. Now the telephone directory does tell you something about London but it is not intended that it should play a major part in a geography lesson. Similarly mathematical proofs were not constructed for the purpose of introducing students to the material but in order to confirm that certain statements about the subject are correct. It may be said in defence of the above that the theory of busprongs consists only of a series of facts and their logical deduction. But if that is the case then there does not seem to be much point in presenting such a logical exercise to students as a course in pure mathematics, and certainly no point in presenting more than one such course, as the logical steps are all much the same.

Keith Austin, in *Educational Studies in Mathematics* 12, 1981, pp. 369–71.

*Manifold*, Spring 1970.

## A new discipline?

Pure Mathematics and Applied Mathatemics.

From a University Prospectus, contributed by A. R. Pargeter.

## Not going steady

In addition, it must be remembered that statistical correlations, however significant, do not necessarily imply a casual relationship.

J. L. Cloudesley-Thompson, *The New Scientist*, 1959, p. 1036. [Per G. N. Copley.] *MG*, 1960, p. 34.

# The Fahrenheit Protractor

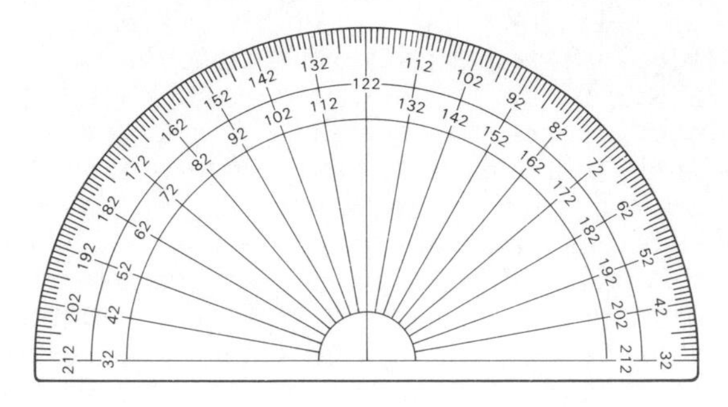


Contributed by W. Wynne Willson, Birmingham.

## A fundamental definition

"Do you know what a mathematician is?" Kelvin once asked a class. He stepped to the board and wrote

$$\int_{-\infty}^{+\infty} e^{-x^2}\,dx = \sqrt{\pi}.$$

Putting his finger on what he had written, he turned to the class "A mathematician is one to whom *that* is as obvious as that twice two makes four is to you."

E. T. Bell, *Men of Mathematics*.

Many things are not accessible to intuition at all, the value of

$$\int_{0}^{+\infty} e^{-x^2}\,dx$$

for instance.

J. E. Littlewood, *A Mathematician's Miscellany*. [Per E. A. Side.] *MG*, 1960, p. 137.

## Random sampling?

And they said everyone to his fellow, come, and let us cast lots, that we may know for whose cause this evil is upon us.

Jonah, 1:7, contributed by Keith Selkirk.

## Mental arithmetic

He (Robert Boyle) deliberately tried to discipline his mind . . . and he found that a very effective way of doing so was to turn to problems of algebra which needed his whole concentration, or to extract square and cube roots in his head. As he was never very good at mathematics he no doubt found this serious occupation a good remedy!

Roger Pilkington, *Robert Boyle, Father of Chemistry*, p. 36. [Per J. Buchanan.] *MG*, 1961, p. 16.

## The snowy tree cricket

Their vibratos are based on repetitions of a single syllable, slowly uttered in a monotonous, melancholy tune . . . throughout the night. Only the tempo varies; but it varies so consistently with changes of temperature that it is known as the "temperature cricket". It is said that if 40 be added to the number of notes per minute divided by four, the result will approximate the number of degrees Fahrenheit.

Blanche Stillson, *Wings, Insects, Birds and Men*. [Per W. H. Cozens.] *MG*, 1961, p. 332.

# Maths is all about us

From a letter to the Editor of the Mathematical Gazette:

Dear Sir,

The recent growth of interest in mathematics seems to be having the effect of increasing the mathematical sophistication of the general public.

The other day at Victoria Railway Station, for example, I saw the following notice displayed by a bootblack:

$$sir + \frac{sir^3}{3!} + \frac{sir^5}{5!} + \ldots ?$$

Pondering the fellow's hyperbolic erudition, I made my way to the bus station, and was pleased to see that my bus had just drawn in. Boarding it, I heard a woman behind me exclaim, in a tone which suggested that the means of solving some problem or other had at that moment dawned on her:

"Get to the back of the CUBE root!"

The emphasis on the penultimate word indicated that until then she had been getting to the back of the square root. For my part, I must confess that while I am able to *extract* square, cube or other roots, the process of "getting to the back of a root" is unknown to me. The woman subsequently boarded the same bus, and I would have asked her to enlighten me, but unfortunately the bus was crowded and I was on an inside seat while she was standing up; moreover she wore a somewhat disgruntled expression.

Alighting from the bus outside the National Portrait Gallery, I was amused to see that among the usual half-dozen or so pavement artists with their coloured landscapes was one who had chalked nothing but these words:

Modern art
Gets worse and worse
So I'm resolved
To write $\tanh^{-1} Q$.

Of course I could not refrain from dropping a coin into his cap and so earning his proffered gratitude.

Yours etc. Basil Mager, Sussex.

***MG*, 1961, p. 250.**

## Interdominational

At the moment there are six upper schools but only three of those are co-educational. Two of the others are single sex and the sixth is ecumenical.

**Spotted by John Backhouse in *The Oxford Times*, 31 May 1985, *MG*, 1986, p. 22.**

## Quirky

Best of all was the school where staff took exception to the QWERTY arrangement and rearranged the keys to read ABCD etc. To their consternation the character on the key which had been hit did not then correspond to what appeared on the screen.

***The Guardian*, 10 April 1986, sent in by David Vincent, *MG*, 1986, p. 206.**

# "L"-plate instructions

We are all familiar with form DL10 of the Ministry of Transport and Civil Aviation which gives a diagram of the distinguishing mark to be displayed on a motor vehicle whilst being driven by a person holding a Provisional Licence. Such "L"-plates can easily be bought from any motor dealer so that very few people will bother about reading the instructions on the official dimensions to be used.

Whilst throwing out a lot of other rubbish one day, I came across this circular and two remarks on it caught my eye. The first of these states:

"NOTE — The diagram reproduced below is HALF ACTUAL SIZE."

Though the correct dimensions to be used are stated in the form "TO BE ? INCHES", I wondered how many people would accept the fact that the given diagram was indeed half actual size, and what interpretation they would put on the word "size" — did it mean that the dimensions of the scale drawing were half those of the plate to be displayed, or that the area of the figure given was half that to be displayed?

A simple mensuration exercise with a reliable ruler (at room temperature) proved the first of these to be wrong. As for the second, I found that most of my non-mathematical friends took this interpretation: I wonder if they know how to solve the problem.

The second remark refers to the appearance of the plate and states:

"The corners of the white ground may be rounded off."

Let us do this by drawing a circle, centre the C.G. of the plate and diameter the width of the plate. We thus have a circular plate. Attacking the corners of the white ground round the "L" itself, we end up with the following:

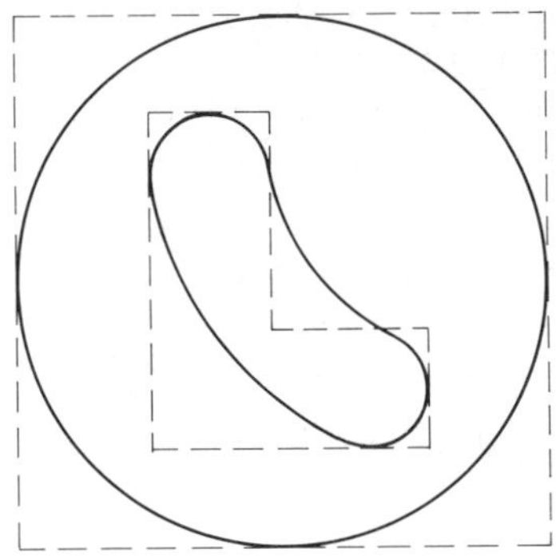

Is it legal, therefore, to hang a red sausage on a white circular plate as the distinguishing mark of a Provisional Licence?

**T. S. Blyth in a letter to *MG*, 1962, p. 67.**

## Imperial jewel

The metric equivalent of the calorie is the kilojoule.

**Carol Bowen, *Cooking for Slimmers*, Sundial Books for Marks and Spencer, 1977, spotted by Brian Head, *MG*, 1986, p. 8.**

## A recurring theme

... the digits 761 to 766 [of $\pi$] are '9999' — a remarkably unrandom sequence so early.

**From the October 1985 edition of *Computing with the Amstrad*. [Per Brian Head.] *MG*, 1986, p. 38.**

## From a mid-nineteenth century exercise book

I saw a tree with tempting fruit
Just sixty-five feet high
But a deep ditch that came between
Forbade me to come nigh

The ditch was fifty-two feet wide
Now I would gladly know
How long a ladder I must get
To reach the topmost bough.

## And a modern response

Pythagoras' will tell you how
With theorem renowned
To reach unto the topmost bough
From down upon the ground.

Assume we have an upright tree
Set upon level ground,
Then at its base 'tis plain to see
A right angle is found.

So first we square the sixty five,
The fifty two the same,
And then the sum of these derive,
Its square root solves our game.

So by Pythagoras' great law
We need a ladder strong,
It must be near three inches more
Than eighty-three feet long!

**From a nineteenth century exercise book in the University Library, Nottingham, with a solution by Keith Selkirk.**

*Manifold*, Autumn 1975.

# No comment

Dear Sir(s),

Import Duties Act, 1958.
Section 6 and paragraph 3 of the Fourth Schedule.

I refer to your application dated 14/2/63 for a duty-free direction in respect of

Mathematical Models

In order that further consideration can be given to your request, would you please state the subject to be taught by this importation.

Yours faithfully,

**Letter received by a grammar school from the Board of Trade, quoted in *The Times Educational Supplement*, 29 March 1963, contributed by A. R. Pargeter.**

# Graphiti

Graphing inequations is a shady business.
Vectors can be 'arrowing.
I like formulae . . . as a rule.
I'm partial to fractions.
Areas leave me flat.
I like angles . . . to a degree.
Maths marks are significant figures.
I could go on and on and on about sequences.
Probability is a chancy business.
Translations are shifty.
Irrational numbers are unrepeatable.
Gradient problems can be a bit steep.
Maths is the limit.
Quadratics are for squares.
Matrices . . . what array to go.
Algebra is $x$-sighting.
Volumes are solid going.
Enlargement grows on you.
Is the square root of *ab* absurd?
Division cuts me up.
Modes are really a common lot.
Sets get it together.
Approximations are exacting.
Fractions are chips off the whole block.
Vectors are going places.
Rotations drive you round the bend.
Complex numbers are unreal.
Average people are mean.
Are statisticians normal?
I feel positive about integers.

**John Hutton, New Zealand, contributed by Andy Begg.**

## Non-empty

On the empty desk sat an empty glass of milk.

**From a Radio 3 broadcast *Close encounters with Kurt Gödel*, as reported by Barry Martin, *MG*, 1986, p. 53.**

## Part of our invisible earnings?

Mathematics counts. Report of the Committee of inquiry into the trading of mathematics.

**From the 1986 Heffer's catalogue, spotted by A. R. G. Burrows, *MG*, 1986, p. 166.**

## Abbott and Costello (1)

Abbott tries to borrow 50 dollars from Costello, who has only 40. So Abbott takes the 40 and says Costello can owe him the 10. When Costello protests, Abbott returns the 40, but then takes the 10 that Costello owes him.

Abbott still wants 50 dollars, so he borrows the 30 that Costello has left, saying that Costello owes him 20. Costello again protests. Abbott returns the 30, but then takes the 20 Costello owes him.

## Abbott and Costello (2)

Costello proves to Abbott that $7 \times 13 = 28$.

First by division; 7 into 2 won't go, 7 into 8 goes 1 and 1 over, bring down the 2 to give 21, 7 into 21 goes 3, completing the 13:

```
7)28(13
   7
  --
  21
```

Next by multiplication, 7 times 3 is 21, 7 times 1 is 7, 1 and 7 is 8.

```
13
 7
--
21
 7
--
28
```

Finally, by addition, count down the right hand column 3-6-9-12-15-18-21-, and then up the left hand column -22-23-24-25-26-27-28.

```
13
13
13
13
13
13
13
--
28
```

**Both contributed by Keith Austin.**

## At the chalkface

*Jan 29 1856:* Breakfasted with Swabey to arrange about teaching in his school [St Aldate's]. We settled that I am to come at 10 on Sunday, and at 2 on Tuesdays and Fridays to teach sums. I gave the first lesson there today, to a class of eight boys, and found it much more pleasant than I had expected. The contrast is very striking between town and country boys; here they are sharp, boisterous, and in the highest spirits — the difficulty of teaching being, not to get an answer, but to prevent all answering at once.

*Feb 1:* The master at St Aldate's School asked if I would join the first class of girls with the boys. I tried it for today, but I do not think they can be kept together as the boys are much the sharpest. This made a class of fifteen: I went on with 'practice' as before.

*Feb 5:* Varied the lesson at the school with a story, introducing a number of sums to be worked out.

*Feb 8:* The school class noisy and inattentive — the novelty of the thing is wearing off, and I find them rather unmanageable.

*Feb 15:* School again noisy and troublesome — I have not yet acquired the arts of keeping order.

*Feb 26:* Class again noisy and inattentive — it is very disheartening, and I almost think I had better give up teaching there again for the present.

*Feb 29:* Left word at the school that I shall not be able to come again for the present. I doubt if I shall try again next term: the good done does not seem worth the time and trouble.

**From the diary of Lewis Carroll, quoted in a letter by Geoffrey Howson, Southampton to *MG*, 1987, p. 147.**

## Statisticians beware

Comparisons are odorous.

**Dogberry in *Much Ado About Nothing*.**
**Contributed by Keith Selkirk.**

## Quotelets from Bibby (*Quotes, Damned Quotes, . . .*)

You have only to take in what you please; to select your own conditions of time and place; to multiply and divide at discretion; and you can pay the National Debt in half an hour. Calculation is nothing but cookery.

**Lord Brougham.**

The overwhelming majority of people have more than the average (mean) number of legs.

**E. Grebnik.**

Statistician's friend: "How is your wife?"

Statistician: "Compared with whom?"

**Anon.**

Statistics can be used to support anything — especially statisticians.

**Franklin P. Jones, *Woman's Realm*.**

Statistics is the only profession which demands the right to make mistakes five per cent of the time.

**Anon.**

Beginners will find that the computer is logical to a disagreeable and intensely frustrating degree.

***OSIRIS II Student Notes*.**

## More quotelets from Bibby (*Quotes, Damned Quotes, ...*)

To call in the statistician after the experiment is done may be no more than asking him to perform a post-mortem examination: he may be able to say what the experiment died of.

**R. A. Fisher (via Eric White).**

In the space of one hundred and seventy six years, the Lower Mississippi has shortened itself two hundred and forty two miles. That is an average of a trifle over one mile and a third per year. Therefore, any calm person, who is not blind or idiotic, can see that in the old oolitic Silurian period, just a million years ago next November, the Lower Mississippi River was upwards of one million three hundred thousand miles long, and stuck out over the Gulf of Mexico like a fishing rod. And by the same token, any person can see that seven hundred forty two years from now the Lower Mississippi will be only a mile and three quarters long, and Cairo and New Orleans will have joined their streets together, and be plodding comfortably along under a single mayor and a mutual board of alderman. There is something fascinating about science. One gets such wholesale returns on conjecture out of such a trifling investment of fact.

**Mark Twain, *Life on the Mississippi*.**

If you don't state your assumptions explicitly, someone else will publish an article doing so for you.

**Anon.**

## Rational approximations: a basic approach

The most familiar quotient in the above sequence is, of course 22/7. As a digression, we quote the following ingenious one-line proof that 22/7 is greater than $\pi$:

$$\int_0^1 \frac{x^4(1-x)^4}{1+x^2}\,dx = \frac{22}{7} - \pi.$$

Who will say that mathematics is devoid of humour?

**G. M. Phillips, *MG*, 1983, p. 247.**

Des MacHale (Cork) points out that an even funnier joke is to try to generalise it to:

$$\int_0^1 \frac{x^{2n}(1-x)^{2n}}{1+x^2}\,dx$$

## The creation of the identity elements

Kronecker said "God made the integers, all the rest is the work of man". It might be added "and the Devil made zero! Multiply by zero and all your work vanishes: divide by zero and it becomes nonsense." Did not Mephistopheles say "Ich bin der Geist der stets verneint" ("I am the spirit of negation") in *Faust?*

**Per Roger North, *MG*, 1961, p. 332.**

*New Scientist.*

## The empty set?

. . . the story told by a fellow editor that occasionally his letters of rejection took the form:

"Dear Sir, Although your paper fills a much-needed gap in the literature, I regret that it cannot be accepted for publication."

**Dr T. H. Osgood, quoted in *Nature*, 9 July 1960. [Per A. P. Rollett.] *MG*, 1964, p. 70.**

## Have you got your factorials right?

$$(2 + 7 = 9!)!$$

***MG*, 1963, p. 25. [Per J. Hind.] *MG*, 1963, p. 159.**

## Not on the National Health!

The opposite angles of a psychiatric quadrilateral are supplementary.

**Howler supplied by Miss E. M. Busbridge. *MG*, 1965, p. 427.**

## Multum in parvo

"There is more sex appeal in a smile from a girl in a summer frock than a dozen in a bikini." (Holiday Hostess.)

*The Daily Mail.* [Per R. E. Green.] *MG*, 1969, p. 345.

## Logic in *The Daily Telegraph*

Never have fewer than 80 per cent. of the electorate in Altrincham and Sale failed to vote.

*The Daily Telegraph*, 1 February 1965.
[Per D. Sculthorpe.] *MG*, 1966, p. 300.

## The Russell paradox again

"This article has seven errors in it. See how many you can spot . . ." (There were six errors in the article that followed.)

Answers: "7. So far there are only six errors; but saying there are seven is itself an error."

*Radio Times*, 7 April 1966. [Per Dr H. M. Cundy.]
*MG*, 1966, p. 300.

## Logic in *The Times*

Sir, — In your third editorial on December 3 you wrote "all scientists are not wise". Did you intend "not all scientists are wise"?

Letter in *The Times*, 9 December 1963.
[Per A. P. Rollet.] *MG*, 1965, p. 427.

## A unique operation

Dear Sir, — During a recent spell in hospital with appendicitis I asked my wife to bring my copy of "The Core of Mathematics" by A. J. Moakes. She did so with the comment that she did not think much of it. On my asking why, she explained that she had glanced through it in order to "take her mind off things". The first thing she saw was the word *operations*. On seeking enlightenment at the back of the book, she found Appendix I, Appendix II, Appendix III, . . .

Yours faithfully, S. F. Hancock, London, SW2.

*MG*, 1967, p. 317.

## There was an old man who said, "Do".

There was an old man who said, "Do
Tell me how I should add two and two?
I think more and more
That it makes about four —
But I fear that is almost too few."

Anon. quoted in *Fantasia Mathematica* by Clifton Fadiman, Simon & Schuster, New York, 1958.

# The 'legal' value of $\pi$

Many people have heard, at some time during their schooling, that 'some legislature, somewhere, once tried to legislate the value of $\pi$, and set it equal to 3.' The very idea of trying to legislate on something as unlegislatable as the value of $\pi$ is ludicrous. But such a bill was actually considered, and the details of this and the way it was handled, culminating in its rejection by the State Senate, can be somewhat amusing.

The bill in question is House Bill No. 246, which was introduced in 1897 into the Indiana State Legislature. The bill was introduced by Representative T. I. Record, representative from Posey County. It does not suggest a single number for the value of $\pi$ but rather suggests several different numbers. The bill was presumably offered as a contribution to education in the state of Indiana.

The first part of the bill states that the area of a circle is equal to the area of a square whose side is $\frac{1}{4}$ the circumference of the circle. If we represent the radius of the circle by $r$, the circumference would be $2\pi r$, and the bill would have us believe that the quantity $(2\pi r/4)^2$ is equal to the area of the circle. We have been taught that the area of the circle was simply $\pi r^2$. The area suggested in the bill would be correct if we assumed that $\pi$ was equal to 4. The bill then goes on to mention that the ratio of the chord to the arc of 90° is as 7 to 8. We would state that the chord of 90° in a circle of radius $r$, was equal to $r\sqrt{2}$, and the arc of 90° is simply $(\pi/2)r$. This latter 'truth' would lead to the value of $\pi = \sqrt{2} \times 16/7$. (This last statement is closer to the truth than was the first statement.) The same paragraph goes on to say that the ratio of the diagonal to one side of a square is as 10 is to 7. The assumption here is that the square root of 2 is exactly equal to 10/7. This approximation is good to 1%. The bill then says that the ratio of the diameter to the circumference of a circle is as 5/4 is to 4 (or, $\pi = 16/5 = 3.2$). The paragraph in question winds up by stating that, "since the rule in present use fails to work, it should be discarded as wholly wanting and misleading in the practical applications." The bill ends with the triumphant statement that the author has "solutions of the trisection of the angle, duplication of the cube, and quadrature of the circle, which will be recognized as problems which have long since been given up by scientific bodies as unsolvable mysteries, and above man's ability to comprehend."

When the bill was first introduced into the House of Representatives, in Indiana, it was referred to the Committee on Swamp Lands. The person who referred the bill to this committee is not known, but if he were known today, he might be honored for having given such a diplomatic appraisal of the worth of the bill.

The Committee on Swamp Lands apparently recognized that the bill was not really in their province, and they recommended that the bill be considered by the Committee on Education. The Committee on Education considered the bill, reported it back to the House, and recommended that it should pass. And it did pass, unanimously, 67 to 0.

In the Senate, the bill fared a little worse. It was referred to the Committee on Temperance! (Perhaps the same shrewd chap who referred the bill to the House Committee on Swamp Lands had a part referring it to the Committee on Temperance. A wonderful choice of Committees!) The bill passed the first reading in the Senate, but this is as far as it ever went. After the first passage, the senators were properly coached, and on the second reading, the Senate threw out this "epoch-making discovery" with much merriment.

***American Scientist, journal of Sigma Xi*, Vol. 53, p. 427A (December 1965), 'The "Legal" Value of $\pi$, and some Related Mathematical Anomalies', by M. H. Greenblatt, quoted in *More Random Walks in Science*, ed. R. L. Weber, Institute of Physics, Bristol, 1982.**

# Note on a theorem of Euler

Euler was great in his own sort of way —
The results that he found were quite good for his day.
But among all his work lies one gem that excels,
That plucks at the heart-strings, the proud spirit quells.
To the mystic, the prophet, the thinker, the sage,
To the white-bearded hermit bent double with age,
Inexpressible solace one formula brought,
Namely $e^{\pi i} + 1 = 0$.

If his spirit is weary, his soul is oppressed,
If he wanders and seeks and he cannot find rest,
If his braces give way, and his trousers fall down,
If he misses his train when he's going to town —
If a dreadful catastrophe thus should befall,
Why, the mathematician will not mind at all.
For the thought will console (as it jolly well ought)
That it's $e^{\pi i} + 1 = 0$.

**Diogenes, source unknown, contributed by A. R. Pargeter.**

## A new angle on conservation laws

Make out of thin cardboard open boxes with the following measurements:

A: 2.4″ × 2.2″ × 2.3″ deep,

B: 2.45″ × 2.25″ × 2.2″ deep,

C: 2.3″ × 2.4″ × 2.1″ deep,

D: 2.35″ × 2.45″ × 2.0″ deep.

B can then used as a lid to A, and D to C. But either complete box will fit inside the other!

**Contributed by A. R. Pargeter.**

# Do we need cubes?

$$\frac{(a+b)^3 + a^3}{(a+b)^3 + b^3} = \frac{a+b+a}{a+b+b} = \frac{2a+b}{a+2b}$$

**Contributed by A. R. Pargeter.**

## A powerful method

$$4^{(2/3)} \times 2^{(-1/3)} = (4 \times 2)^{(2/3 - 1/3)} = 8^{(1/3)} = 2$$

**Contributed by A. R. Pargeter.**

## It all adds up

A pair of snakes were having difficulty in producing a family. They went along to the family planning clinic and related their problem. "We would like to have babies but we can't multiply, we're only adders", they said. They were given counselling, and after a few sessions went away happy, and in due course a family of bouncing young adders appeared. "How did you manage it?" asked Mrs Adder's best friend in a particularly friendly moment. "Oh, it was very easy", she replied, "They told us to cut down some trees and make log tables!"

**Contributed by Valerie de la Praudière, Winchester.**

## Positively or negatively skewed?

. . . If the standard of driving on motorways justified it, then I think a case for raising the speed limit would exist. As it is, surveys have shown that over 85 per cent of drivers consider themselves to be above average! In reality, the reverse is more likely to be true.

**From a letter to *Motorcycle Rider*, Number 13, contributed by J. R. Bradshaw, Tamworth.**

## $x = +?$

$$(5 - 3x)(7 - 2x) = (11 - 6x)(3 - x)$$
$$\Rightarrow 5 - 3x + 7 - 2x = 11 - 6x + 3 - x$$
$$\Rightarrow x = 1.$$

**Contributed by A. R. Pargeter.**

## Up you too!

A Roman epic film was issued with an 'X' certificate. As it was too long it was shown in two parts, each having a 'V' certificate.

**Contributed by Keith Austin.**

# OCCAM'S BARBER

*Manifold.*

## The best of Colemanballs

| | |
|---|---|
| . . . and Dusty Hare kicked 19 of the 17 points. | *David Coleman.* |
| Bill Tindall has done a bit of mental arithmetic with a calculator. | *Ted Lowe.* |
| The dart is almost diametrically in the middle of the treble twenty. | *David Lanning.* |
| 53% of all cars coming into Britain for the first time are imported. | *Monty Modlyn.* |
| There's one minute sixty seconds to go. | *Eddie Waring.* |
| He made a break of 98 which is almost one hundred. | *Alan Weekes.* |
| Oh, that cross court angle was so acute is just doesn't exist. | *Dan Maskell.* |
| So that's 57 runs needed by Hampshire in 11 overs, and it doesn't need a calculator to tell us that the run rate required is 5.1818 recurring. | *Norman Demesquita.* |
| . . . and with eight minutes left the game could be won or lost in the next five or ten minutes. | *Jimmy Armfield.* |
| And Ritchie has scored 11 goals, exactly double the number he scored last season. | *Alan Parry.* |
| Attached to the bottom corner is a rope of infinite length. | *Gordon Burns.* |
| . . . and it's exactly nine minutes past nine — and that doesn't happen very often. | *Tim Boswell.* |
| In those days, number two was in a funny sort of way, also number one. | *Jimmy Saville.* |

**From *Colemanballs*, Vols. 1 and 2, originally in *Private Eye* and reproduced by permission.**
**Contributed by Peter Jones, Nottingham.**

# Mr C. V. Durell's "Genarith"

This prolific writer of mystery books — it would be wrong to call them "thrillers" — fully maintains his reputation in the work before us,* and the reader who delights in problems will find here a plenty. Who are the characters? How many of them are there? What is the plot of the story? These are some of the questions which the mysteriologist will approach with a simple interest which will intensify in proportion to his success in discounting the false clues with which the volume abounds.

Mr Durell has adopted a somewhat statistical and episodic style which does not make for easy reading; each incident is classified and numbered and presented in a completely objective and baldly matter-of-fact form in the minimum of words. The author's descriptive style is frankly lamentable and one feels that he has not succeeded in making his characters live. He veils them in an anonymity which cuts across the reader's sympathy, referring to them impersonally as "a man" or as "A, B and C"; thus it is not easy always to relate a character to the incidents of his career, more especially since the author gives no clue to the thought-processes or motives leading up to the events he narrates. He has, too, a disconcerting habit of suddenly shifting the locale of his story without the slightest warning. Thus on p. 141 an explorer investigating the sources of the Amazon appears to have been transported thither from the Ganges in the twinkling of an eye, and on p. 300 he gets himself unaccountably mixed up with Feddens and qirats in Egypt. There is similarly the Home Office inspector who travels vaguely round the country enquiring into the Local Government of unsuspecting towns, with especial reference to their finances — p. 142 Winchester, p. 268 Oxford, p. 386 Leeds, p. 388 Poole. These people certainly get about!

The main action of the book takes place in this country and in France, as we may deduce from the frequency with which the characters are called on to exchange English money into French and vice versa; the period was evidently before the war, since cigarettes were 20 for 1/- [5p] (p. 27). As might be expected, the most clearly defined personality is the villain. He begins in a small way as a barrow-boy, putting only 12 peaches in boxes ostensibly containing 18 (p. 30) and selling oranges two-ninths of which are bad (p. 42); the absence of any mention of nylons confirms our estimate of the period. Considering the company he keeps — in one chapter the words "common", "vulgar" and "improper" recur monotonously — how can he be expected to remain "integer vitae, scelerisque purus"? From his first petty defalcations he is able to make enough to buy a small milk round; he waters the milk and narrowly escapes well-deserved ruin when he spills some of it (p. 175). Even when he plays cricket he cannot refrain from falsifying his average (p. 236), and he advertises a car for sale claiming for it a performance far in excess of the truth (p. 116). For a time his career is obscure. There is a cryptic reference to a Derby run in record time (p. 237) on which he presumably wins a packet, since he gives a celebration dinner but fails to pay nearly half the bill (p. 256) though he spends £96 on shirts (p. 116). In an access of remorse he pays £1195 as conscience money to the Exchequer (p. 258), but he ends miserably with an overwhelmingly number of bankruptcy orders against him (p. 142).

A less clearly drawn character is a man who is Something in the City. His father encouraged him to save in his boyhood (p. 157), and later in his life he was able to do a little money-lending at $12\frac{1}{2}\%$ (p. 152). He has a setback when a bank holding some of his savings fails (p. 158), and after dabbling in insurance (p. 165) and bill discounting (p. 221) he loses money in two speculations obviously suggested by the villain (p. 257). But he is not

*Clement V. Durell, *General Arithmetic for Schools*, Bell, 21st impression, 1951, pp. xvi, 441.

dismayed: some lucky deals in foreign stocks (p. 366) and a profitable foreign exchange operation (p. 398) set him on his feet. Only once does he fall foul of the law, when he is trapped by a policeman for travelling at 68 m.p.h. in a built-up area (p. 378).

We can do no more than mention the scientist with his germ-cultures (p. 147), the swimming enthusiast, who is always building swimming baths and emptying and refilling them (*passim*) or the College Youths whose sense of timing is poor (p. 13); but attention must be drawn to the book's most striking feature — the complete absence of love interest. On p. 50 there is certainly mention of a co-educational school where doubtless began a boy-and-girl romance which culminated in the marriage (on p. 51) of a man of 27 to a woman of 24, but there is not a word about the courtship nor of the outcome of the marriage except that it lasted 54 years and a hint that the wife joined a committee (p. 265); perhaps it was that Road Safety Committee which discovered that if you don't want to be killed on the roads you should not be a child under ten (p. 102).

A remarkable book: and if you think this review is sketchy see if you can do any better yourself!

**B. A. Swinden, *MG*, 1955, p. 214.**

## Two plus two

Someone asked an accountant, a mathematician, an engineer, a statistician and an actuary how much 2 plus 2 was. The accountant said "4". The mathematician said "It all depends on your number base". The engineer took out his slide rule and said "approximately 3.99". The statistician consulted his tables and said "I am 95% confident that it lies between 3.95 and 4.05". The actuary said "What do you want it to add up to?"

**Anon.**

***Manifold*, Spring 1973.**

## Quotelets from Bibby (*Quotes, Damned Quotes, . . .*)

If you torture the data long enough, it will confess.

**Ronald Coase.**

If at first you don't succeed, you're just about average.

**Bill Cosby (American comedian).**

It is my earnest hope never to read "lies, damned lies, and statistics" more than once a month, it long since having lost any excitement to me.

**Frank Oliver.**

Variance is what any two statisticians are at.

**Anon.**

It's like the tale of the roadside merchant who was asked to explain how he could sell rabbit sandwiches so cheap. "Well" he explained, "I have to put some horse meat in too. But I mix them 50–50. One horse, one rabbit."

**Darrell Huff, *How to Lie with Statistics*.**

The Dirty Data Theorem states that "real world" data tend to come from bizarre and unspecifiable distributions of highly correlated variables and have unequal sample sizes, missing data points, non-independent observations, and an indeterminate number of inaccurately recorded values.

**Anon.**

## Idle chatter

The law of conversation of momentum must be applied in vector form when solving Problems 186 189.

***Problems in Undergraduate Physics: I, Mechanics,* S. P. Strelkov and I. A. Yakovlev, translated by D. E. Brown, Pergamon, 1965. [Per Dr J. H. Wilkinson.]**

(A new use for those moments of inertia, I suppose. Ed.)

***MG*, 1968, p. 218.**

## Integral number theory

Announcing a new journal:

*Journal of Number Theory*

. . . Each volume . . . will contain four numbers.

**[Per Dr A. F. Ruston.] *MG*, 1970, p. 336.**

## A possible use for your non-returnable Klein bottles

Poincaré himself . . . made some mistakes. For example, (he) stated that there could not be a closed 'one-sided' surface. Yet (the Klein bottle), similar to a sort of fly trap in use in France, is an instance of a surface of such a kind.

**From J. Hadamard, *Later Scientific Work of Henri Poincaré*. (Rice Institute Pamphlet, 1933). [Per Dr A. G. Howson.] *MG*, 1972, p. 233.**

# Clean round the bend

N10. (1) Any drain or private sewer shall . . . (d) be laid in a straight line between points where changes of direction or gradient occur.

**From *The Building Regulations 1972* (HMSO). [Per Brian Milo.] *MG*, 1976, p. 150.**

# Warning to Torricelli!

If that which is in a hollow vessel were taken out of it without anything entering to fill its place, the sides of the vessel having nothing between them would be in contact.

**Descartes. [Per G. S. Light.] *MG*, 1975, p. 60.**

# Here's all you do:

Simply melt butter (or margarine) in a saucepan. Add your 24 Pascall Marshmallows and cook over a low heat. Stir constantly until marshmallows are melted. Remove from heat. Add your 5 cups of Kellogg's Rice Bubbles. Press the warm mixture into buttered 9″ × 9″ cake tin. Add cherries if desired. When cool, cut into pieces $1\frac{3}{4}'' \times 1\frac{1}{2}''$. Makes 35 tempting treats.

**From an Australian cereal packet. [Per Professor J. C. Burns.] *MG*, 1972, p. 233.**

# Jobs for re-deployed mathematicians?

*Drug Problems and their Management*. Derek Richter, published for the Association for the Prevention of Addition, 1971.

**From a Department of Education and Science booklet, *Drugs and the Schools, MG*, 1974, p. 77.**

***Manifold*, Winter 1972.**

## Improve your golf

The extra distance is obtained through added club-head speed. This is gained by the lighter shaft giving a lower overall weight — 12 oz instead of $13\frac{1}{2}$ oz in a driver — and enabling weight to be transferred to the club-head. It adds up to the formula: Force equals weight + velocity.

*The Sunday Express*, 24 June 1973. [Per D. R. Brown.] *MG*, 1974, p. 264.

## I do and I understand?

We assert that if the resistance of the air be withdrawn a sovereign and a feather will fall through equal spaces in equal times. Very great credit is due to the person who first imagined the well-known experiment to illustrate this; but it is not obvious what is the special benefit now gained by seeing a lecturer repeat the process. It may be said that a boy takes more interest in the matter by seeing for himself, or by performing for himself, that is by working the handle of the air-pump: this we admit, while we continue to doubt the educational value of the transaction. The boy would also probably take much more interest in foot-ball than in Latin grammar; but the measure of his interest is not identical with that of the importance of the subjects. It may be said that the fact makes a stronger impression on the boy through the medium of his sight, that he believes it more confidently. I say that this ought not to be the case. If he does not believe the statement of his tutor — probably a clergyman of mature knowledge, and a blameless character — his suspicion is irrational, and manifests a want of the power of appreciating evidence, a want fatal to his success in that branch of science which he is supposed to be cultivating.

Isaac Todhunter, an Oxford Professor of Mathematics, in *The Conflict of Studies and Other Essays*, Macmillan, 1873, and quoted in *A Random Walk in Science*, ed. R. L. Weber, Institute of Physics, Bristol, 1973.

## It's all bull

Big Chief Sitting Bull had three wives, as was the custom of the tribe. Falling Leaf was the youngest and most beautiful wife, and weighed just six stones. The middle wife was Running Nose, who weighed nine stones; but the oldest wife, Rolling Steam, weighed a massive fifteen stones. One Christmas Sitting Bull bought each of his wives a new skin to sit on — a deer skin for Falling Leaf, a buffalo skin for Running Nose, and a rug made of hippopotamus skin for Rolling Steam. The Medicine Man of the tribe thereupon recorded in his diary the fact, now well known, that the squaw on the hippopotamus was equal to the sum of the squaws on the other two hides.

Contributed by John Deft, Bristol, and others in various other versions.

## The long way round

Boeing explains that these are marked in centimetres of depth and then first converted into litres, then pounds, and finally kilograms.

***The Sunday Times*, 31 July 1983. [Per Hamish Sloane.] *MG*, 1984, p. 287.**

## Addition

One and one make two,
But if one and one should marry,
Isn't it queer —
Within a year
There's two and one to carry.

**Author unknown, contributed by John Deft, Bristol.**

## A mnemonic

Mnemonics are commonly an appeal against logical principles. The following is intended as an appeal against arbitrary rules to logical principles. Perhaps it is not too flippant for publication.

"Please, Sir; is his income 5 per cent of the Stock he bought or of the price he paid for it?"

The railway neither knows nor cares
How you obtained your blessed shares,
Bought in a pawn-shop second-hand
Or found in Joe's umbrella-stand.
Why should it pay you 5 per cent
On what *you say* that you have spent?
And if you stole them, I'm afraid
Your fortune won't be quickly made
By 5 per cent of *what you paid*.

**'Elpis' in *Mathematical Gazette*, 1924, reprinted 1971, p. 223.**

## A bell by any other name would smell as sweet

Sometimes the normal distribution is called Gaussian, especially in engineering and physics. In France it is called Laplacean. These names are probably used because the distribution was invented by de Moivre.

**F. Mosteller, R. E. K. Rourke and G. B. Thomas, *Probability with Statistical Applications*, Addison Wesley, 1970. [Per Eric Door.] *MG*, 1976, p. 193.**

## Definition by example

A scientist, a logician and a mathematician were driving through a remote area when they saw a single sheep which was black. The scientist said, "The sheep round here are black." The logician said, "Some of the sheep round here are black." The mathematician said, "There is at least one sheep round here which is black on at least one side."

**Contributed by Keith Austin.**

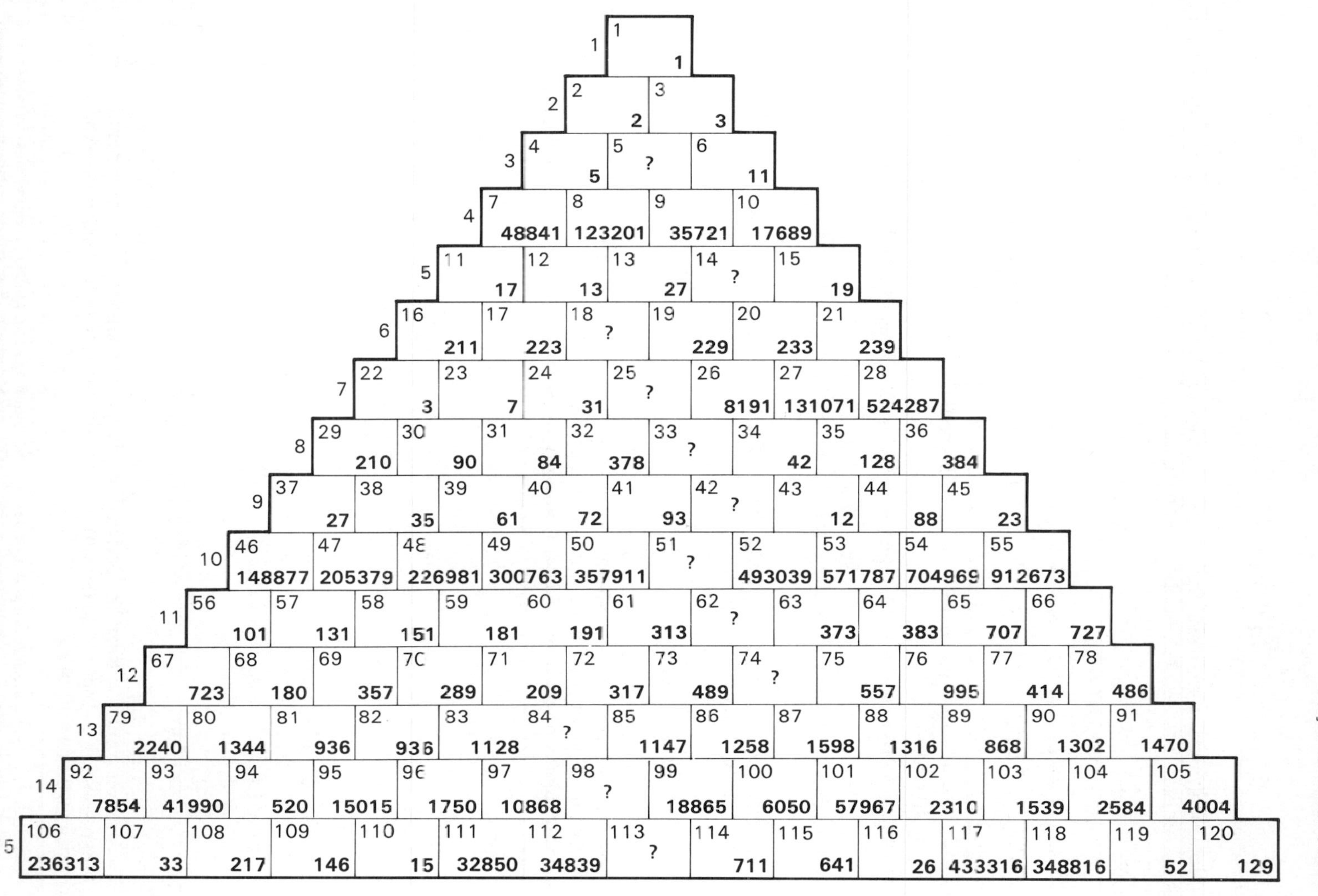


[For a hint see p. 58.]

Contributed by A. R. Pargeter.

# Spherical aberration?

The following problem was set in a problem-solving competition at the Royal Wolverhampton school and has been sent to us by J. N. MacNeill.

> In Fatland it is considered physically beautiful to be the shape of a snooker ball, that is, to be spherical.
>
> "Spherical is beautiful," said Helen, eating her spaghetti sandwich.
>
> "I'm more spherical than you are," said Julia, eating her potato risotto.
>
> "Don't eat with your mouths full," said someone else. "I mean don't speak with your mouths full. In any case you're either spherical or you're not; you can't be *more* spherical."
>
> "Well . . ." Julia thought. "I'm *nearer* to being spherical than Helen is."
>
> "No you're not," countered Helen, "because my circumference is bigger than your circumference!" [In Fatland, the word 'waist' is almost obscene.]
>
> "I agree that you're bigger than me all round," said Julia with a tinge of envy, "but we're not talking about *size*, we're talking about *shape*, and my shape is more like a sphere than yours is!"
>
> "How can you say that? How can you tell," demanded Helen, "if one shape is more like a sphere than another is?"
>
> Helen's question is a good one. Invent a way of deciding which of two given solid shapes is nearer to being spherical. It does not matter if your method would be difficult to use in the case of complicated shapes like Julia and Helen. Any resemblance to any living person is entirely to be expected.

## Solution

Richard Dobbs (Magdalen College, Oxford) writes:

> I would suggest a flatness factor given by
>
> $$\frac{360\pi(\text{volume})^2}{(\text{surface area})^3}$$
>
> where ten would be their Bo Derick! This is independent of size and dependent on shape only. it will be a maximum for a sphere as a sphere has maximum volume for a given surface area.

***Mathematical Spectrum*, Vol. 17, No. 3 and Vol. 18, No. 2, reproduced by permission.**
**The word 'flatness' is, of course, a misprint for 'fatness'.**

## Some slogans

Geometry keeps you in shape.

Calculating people go places.

Decimals make the point.

Maths counts. (Should we apologise to Cockcroft?)

Sum people are wonderful.

Mathematicians: how do you measure up?

Get factorised!

Einstein was ahead of his time.

Archimedes was wet.

Newton started with an Apple (*sic*).

Lobachevski was out of line.

Absolute zero rules 0 K. (The degree symbol should not be used with the Kelvin scale).

**Contributed by Andy Begg, New Zealand.**

*"Forceps . . . scalpel . . . pocket calculator . . ."*

***The Guardian.***

## Small audiences . . .

It was commonly assumed at the BBC and TV-am that most people would watch [breakfast television] for a maximum of 0 minutes before heading to work.

***The Observer*, 23 January 1983. [Per F. H. C. Oates.] *MG*, 1984, p. 91.**

## . . . and large

A total television audience of 10 billion, more than twice the world population, watched the year's World Cup Final.

***The Times*, 18 December 1982. [Per B. Cook.] *MG*, 1984, p. 99.**

## If $a = b$ and $b = c$, then $a = c$

A lazy dog is a slow pup.
A slope up is an inclined plane.
An ink-lined plane is a sheet of writing paper.
Therefore a lazy dog is a sheet of writing paper.

**From *Mathematics and Humour*, ed. A. Vinik and L. Silvey, National Council of Teachers of Mathematics, Reston, Virginia, and reproduced by permission.**

## Research ad nauseam

A division of Glaxo which produces a travel sickness cure has just announced: 'Research has shown that the 15 tablets in the current pack are excessive, so to reduce wastage this has been reduced to 12 tablets.' And, by the way, it adds, the price will stay the same. Research will no doubt show that a 20% increase in profit is not a bit excessive.

***The Daily Telegraph*, 5 April 1983. [Per A. R. Pargeter.] *MG*, 1985, p. 110.**

## Honesty is the best policy

The book falls into two parts.

**From a publisher's advertisement for *Modular Programming*. [Per David Green.] *MG*, 1976, p. 240.**

## Permutations?

Sir, — As a GCE "O" level examiner, perturbed by recent correspondence, I now realise that to collect misspellings of the word "isosceles" is indulgence in a blood sport. This year's batch of innocents in whose slaughter I have been an accessory produced the following variations. Bogey for a batch of 500 scripts may be taken as 25:

> Isoscles, isoscoles, isosoles, isoseeles, isoselles, isoseles, isosecles, isosoceles, isoseceles, isosolees, isosolese, isoceles, isocles, isocelles, issocolese, issoceles, issoleles, iscoseles, iscoscelese, iscoceles, iscosceles, iscoles, iscosoles, iscocolles.

The isosceles triangle is a vested interest. It must go. — Yours faithfully, E. G. Bowley.

**From a letter to *The Guardian*, 15 July 1964, reproduced by permission. Contributed by A. R. Pargeter who adds 'Has it ever occurred to you that we could get along very well without the word anyway?'**

## One map does not a theorem make

The object of this book is to discuss some of the amazingly powerful and diverse mathematical ideas . . . that were developed during the hundred year assault on the problem of producing a four colour map of the world.

**From a Wiley booklist, noted by Robin Wilson. *MG*, 1985, p. 294.**

## A new twist on infinity

That queer quantity "infinity" is the very mischief, and no rational physicist should have anything to do with it. Perhaps that is why mathematicians represent it by a sign that is like a love-knot.

**Sir Arthur Eddington, *New Pathways in Science*, quoted in *Fantasia Mathematica* by Clifton Fadiman, Simon & Schuster, New York, 1958.**

## Non-Euclidean London

Though the Tower looks square it is rectangular, and three of its corners are not right angles.

***The Illustrated London News*, Christmas number, 1976. [Per the Rev. B. V. Lagrue.] *MG*, 1977, p. 272.**

## Kai Lung scrambles his eggs

Either shell or unshell the [6] eggs by knocking one against the other in any order. Since, when two eggs collide, only one of them will break, it will be necessary to use a seventh egg with which to break the sixth. If, as it may very well happen, the seventh egg breaks first instead of the sixth, an expedient will be simply to use the seventh one and put away the sixth. An alternative procedure is to delay your numbering system and define that egg as the sixth egg which breaks after the fifth egg.

**From Buwei Yang Chao, *How to cook and eat Chinese*. [Per A. R. Pargeter and J. Glenn.] *MG*, 1978, p. 152.**

## A case for Dr Venn

Sir, — I read a report recently which stated that a certain large and above average percentage of women who are below age 25 at marriage are pregnant. I was therefore surprised to read that although 'a third of women who marry are pregnant' only 'a sixth of babies born in wedlock to mothers under 25 are conceived before marriage.'

However if the statistics are correct and a third of women are pregnant at marriage, how can three-quarters of married couples remain 'childless after two years'? . . .

By the way, did you realise that 50 per cent of all married couples are women?

**From a letter to *The Guardian*, 6 January 1978. [Per C. D. J. Mills.] *MG*, 1979, p. 50.**

*Manifold*, Summer 1969.

# Was your investment cost-effective?

Dear Member,

A FURTHER INVITATION

The value of calculating machines as a teaching aid at all levels, is now widely accepted by mathematics teachers. It may not be so widely known however, that, owing to pioneering work over the past ten years, Britain leads the world in the use of this method of teaching.

The Brunsviga 13 RM at £39.10.0 [£39.50] educational price (or less for bulk purchase) with 10 × 8 × 13 capacity, full tens transmission, one-handed operation and back-transfer, is the latest of the family of Brunsvigas well known for 75 years.

The large number of Brunsvigas already in use in schools all over the country, is a testimonial to the excellent value which this robust, low-cost, machine represents.

There are still, however, many schools with very few calculating machines and even entirely without. This is why we have pleasure in once again inviting members to try, without obligation, a Brunsviga 13 RM for a fortnight, on receiving a request on school letter-heading.

Yours very truly;
EDUCATIONAL ADVISER.
Brunsviga Calculator Division.

**This advertisement appeared in *MG* in 1968.**

# A new electronic age

The last year or so has seen a rapid development in electronic calculators. Two factors — advancing technology and volume production — have made machines both smaller and cheaper. It is now possible to buy an electronic calculator, which will add, subtract, multiply and divide instantly at the touch of a button, for a little under £40: the small advertisements in the week-end newspapers are full of them. They are portable too — even pocket sized battery models are available. There are forecasts that by 1982 the cost could be as low as £10.

**Editorial, *Mathematics in School*, July 1972.**

## Remembering pi

[Many rhymes have been composed giving the value of $\pi$ by the number of letters in each word. Here is one.]

Sir, I bear a rhyme excelling,
In mystic force and magic spelling,
Celestial sprites elucidate,
All my striving can't relate.

**Anon., quoted in a letter from Dr S. Roy to *The Guardian*, 15 March 1984. Contributed by D. Brown.**

## . . . or take an umbrella

When caught in the rain without mack,
Move as fast as the wind at your back.
But if the wind's in your face
The optimal pace
Is as fast as your legs can make track.

**D. Brown, York, inspired by an article in *MG*, October 1976.**

## Decent functions

"... a decent function. What do I mean by a decent function? Well, a decent function is a function which possesses all the properties which it would be assumed to possess by any gentleman."

**Professor A. S. Besicovitch, Trinity College, Cambridge, during a lecture. Contributed by A. R. Pargeter.**

## 'A lotta bottle'

If we bought milk by the half litre, how many bottles would you have to buy to have roughly four pints? Answer: $4\frac{1}{2}$.

***Help Your Child with Maths.* [Per Peter Ransom.] *MG*, 1985, p. 303.**

## The area of a trapezium

If a trapezium has parallel sides $c$ and $d$ where $c > d$, $a$ and $b$ are the other two sides and $s = \frac{1}{2}(a + b + c + d)$, then its area is:

$$\frac{c+d}{c-d}\sqrt{\{s(s-a)(s-b)(s-c+d)\}}$$

**Contributed by A. R. Pargeter.**

## Quadratics

$$(x + 3)(2 - x) = 4$$

$$\Rightarrow x + 3 = 4 \text{ or } 2 - x = 4$$

$$\Rightarrow x = 1 \text{ or } x = -2$$

[Note: $(x - \alpha)(x - \beta) = 0$ can always be "solved" by this method if written:

$$(x + 1 - \alpha)(1 + \beta - x) = 1 - \alpha + \beta]$$

**Contributed by A. R. Pargeter.**

# An obtuse angle

Find the largest angle of a triangle with sides 4, 7 and 9 units:

$$\sin C = (9/7) = 1.2857,$$
$$1 = \sin 90°$$
$$0.2857 = \sin 16° \, 36'$$
$$\Rightarrow C = 106° \, 36'$$

[Note: will it always work with a triangle whose sides are $a$, $a + b$ and $a + c$ where $a^2 + b^2 = c^2$?]

**Contributed by A. R. Pargeter.**

# Wot no factors?

$$\frac{a^2 - b^2}{a - b} = a + b$$

[Reasoning: $a$ into $a^2$ goes $a$, $b$ into $b^2$ goes $b$, two minuses make a plus.]

**Contributed by A. R. Pargeter.**

# What's the problem?

A Möbus strip may easily be untwisted — without its ends being separated — in 4-dimensional space.

**W. B. Phillips, *Physics for Society*. [Per R. P. Boas.] *MG*, 1978, p. 45.**

# Matters of degree

Leonardo of Pisa was also associated with the Rabbitt Breading problem. [Our correspondent adds that this is thought by scholars to be an early form of the Ham Sandwich Theorem.]

**From a B.Sc. finals History of Mathematics script. [Per Geoffrey Howson.] *MG*, 1978, p. 55.**

# Horse sense

Once there was a supersmart horse that could add, subtract, multiply, divide and even extract roots. Someone suggested that the horse should try a bit of analytic geometry (ordered pairs, graphing, and so on). The horse promptly died. *Moral: You can't put Descartes before the horse.*

**From *Mathematics and Humour*, ed. A. Vinik and L. Silvey, National Council of Teachers of Mathematics, Reston, Virginia, 1978, and reproduced by permission.**

# Higher maths

Examiners for the new GCSE science papers in the East Midlands won't be in any doubt as to how the papers should be marked. Thanks to the regional examinations board's latest guidelines, practical science skills will be a doddle to assess. They will be marked on a six point scale 0–5, and just in case any of the examiners can't recall any of their own childhood lessons, the board helpfully adds the following tips: "A mark of 0 would be a performance below that of mark 1. A mark of 2 would be a performance better than mark 1, but not worthy of 3, and similarly a mark of 4 would be a performance of mark 3 but not worthy of mark 5." Well that adds up.

**From *The Times Diary*, 10 November 1987, by PHS and reproduced by permission. Contributed by Charles Cooke, Nottingham.**

## A gem from Whitehall

New method of calculating concessionary fares for pensioners. "Number of pensioners multiplied by the square root of predicted service miles per hectare, where predicted service vehicle miles is generated by a regression of actual vehicle miles against numbers of persons in households without access to a car." Michael Brereton, council leader at Newcastle-under-Lyme, insists improbably: "We have been campaigning for this for years."

**From an unsourced and undated newspaper cutting.**

## Extracted from a review (unpublished)

By the then Editor.

There is much that he does not say that he means, that he knows you know he means, and so you cannot contradict what he does not say, what you know he means to say — and yet you cannot agree with what he does say, for you know that that will be taken to mean that you agree with what he does not say as well, and to the latter you are firmly opposed. I hope the reader will survive this. That is how I felt after reading 'X'.

***MG*, 1912, reprinted 1971, p. 220.**

## Quotelets from Bibby (*Quotes, Damned Quotes, . . .*)

There is a special department of hell for students of probability. In this department are many typewriters and many monkeys. Every time that a monkey walks on a typewriter, it types by chance one of Shakespeare's sonnets.

**Bertrand Russell, *Nightmares of Eminent Persons*.**

Lies, damned lies and Sun exclusives . . .

***Daily Mirror* headline, 22 October 1982.**

Statistics means not ever having to say you're certain.

**Myles Hollander, after Erich Segal's *Love means not ever having to say you're sorry*.**

In earlier times, they had no statistics, and so they had to fall back on lies.

**Stephen Leacock (1869–1944).**

We look forward to the day when everyone will receive more than the average wage.

**Australian Minister of Labour, 1973.**

For so it is, O Lord, I measure it; But what it is I measure I do not know.

**St Augustine.**

## Intuition

I have had my solutions for a long time, but I do not yet know how I am to arrive at them. (Gauss)

It's plain to me by the fountain that I Draw from, though I will not undertake to prove it to others. (Newton).

**Quoted in W. M. Priestly, *Calculus: an historical approach*. [Per E. H. H. Lockwood.] *MG*, 1981, p. 33.**

*"I didn't understand all that stuff he said between 'Good Morning, Class' and 'That concludes my lecture for today'."*

# The art of finding the right graph paper to get a straight line

As any fool can plainly see, a straight line is the shortest distance between two points. If, as is frequently the case, point A is where you are and point B is research money, it is most important to see to it that the line is as straight as possible. Besides, it looks more scientific. That is why graph paper was invented.

The first invention was simple graph paper, which popularised the straight line (figure 1). But people who had been working the constantly accelerating or decelerating paper had to switch to log paper (figure 2). If both coordinates were logarithmic, log-log paper was necessary (figure 3).

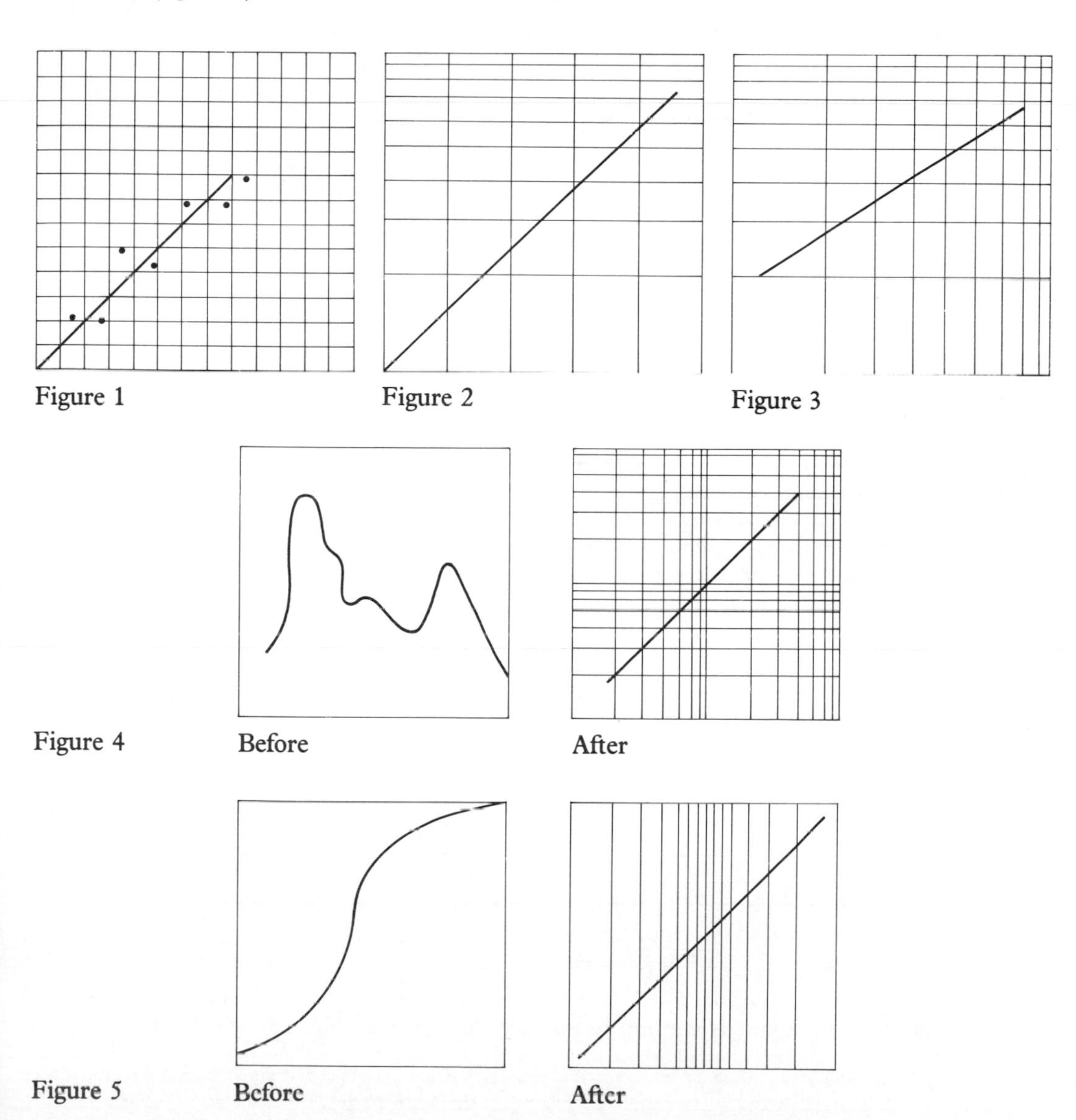

Figure 1 Figure 2 Figure 3

Figure 4 Before After

Figure 5 Before After

Or, if you had a really galloping variable on your hands, double log–log paper was the thing. And so on for all combinations and permutations of the above (figure 4).

For the statistician, there is always probability paper, which will turn a normal ogive into a straight line or a normal curve into a tent. It is especially popular with statisticians, since it makes their work look precise (figure 5).

Sometimes correlation coefficient scattergrams come out at 0.00 with a distribution shaped like a matzo bull (figure 6A). But using 'correlation paper' Pearson's *r*'s of any desirable degree of magnitude can be obtained (figure 6B). Naturally, negative correlation paper is available; it simply points the diagram the other way.

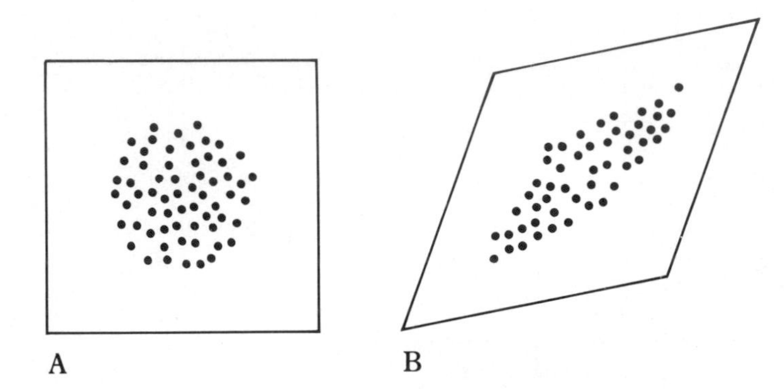

Figure 6 A B

When you get a cycle where you should be getting a straight line, you use the following method. First, the peaks and troughs of the original plot are marked (figure 7A). Then, an overlay of transparent plastic sheet is put over it, and the dots alone copied. Now, it is obvious that these points are simply departures from a straight line, which is presented in dashed form (figure 7B). Finally, the straight line alone is recopied on to another graph paper (figure 7C).

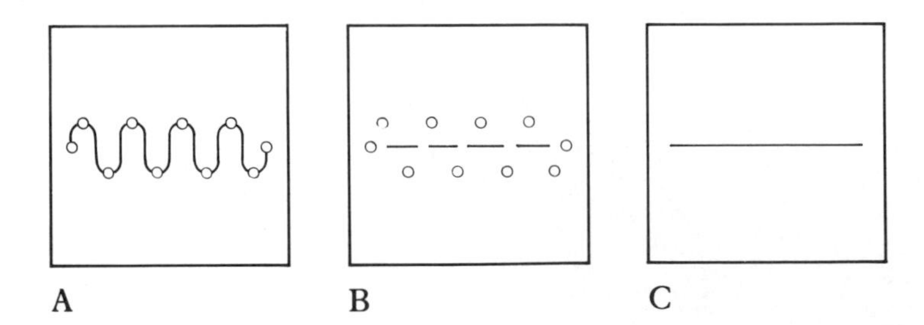

Figure 7 A B C

There is nothing so graphic as a graph to make a point graphically.

S. A. Rudin, Condensed from *Journal of Irreproducible Results*, 12, No. 3 (1964) and taken from *A Random Walk in Science*, ed. R. L. Weber, Institute of Physics, Bristol, 1973.

*"I think you should be more explicit here in step two."*

**Contributed by Stig Olsson.**

## A conservative estimate

. . . between 10 and 20 000 Cubans in Ethiopia . . .

***The Daily Telegraph*, 25 February 1978.**
**[Per F. J. Budden.] *MG*, 1979, p. 175.**

## United we stand

The four normal channels were reduced to one — dubbed unofficially BBC-10 as the product of Radios One, Two, Three and Four.

***The Guardian*, 23 December 1978. [Per Eric Primrose and D. G. H. B. Lloyd.] *MG*, 1980, p. 106.**

## 007

Once is happenstance, twice is coincidence, any more and its enemy action.

**James Bond (Ian Fleming).**

## For mathematicians who shine

$4n\eta$ ft. [Four n eta feet].

**Slogan from the shoe-cleaning cloths at the last British Mathematical Colloquium.**
**[Per A. K. Austin and others.] *MG*, 1981, p. 78.**

# Oh, G!

*The scene is a conference room. Several scientific and mathematical terms can be seen, seated around a log table and talking excitedly. Some shades of noted mathematicians etc. are in the room. Evidently a debate at a company board meeting is about to begin; the Force of Gravity enters, and takes the chair.*

FORCE OF GRAVITY: Will Time please read the minutes?

*(Time does so, but no one is very interested. The noise increases, until there is an outburst by Resultant Force.)*

RESULTANT FORCE: Motion! I want the motion!

KINEMATICS, DYNAMICS, ETC.: We want the motion!

VELOCITY: Please get a move on.

MINIMUM: Please do, I've only the shortest possible time.

MAXIMUM: What's the rush? Let's make the most of it.

FORCE OF GRAVITY: Very well, we will begin the debate. The motion is . . .

FRICTION *[interrupting]*: I oppose it.

FORCE OF GRAVITY: But you haven't heard what it is yet.

FRICTION: I know, but I invariably oppose the motion.

THE SHADE OF ARCHIMEDES: Eureka! A man of principle!

FRICTION: I am not a$\mu$used.

FORCE OF GRAVITY: To proceed. The motion is — that in order to simplify calculations . . .

ACCELERATION: Faster!

FORCE OF GRAVITY *[continuing unperturbed]*: . . . the value of $g$ to be changed from 981 cm/sec$^2$ to 1000 cm/sec$^2$.

SI UNITS: You're behind the times. You mean from 9.8 m s$^{-2}$ to 10 m s$^{-2}$.

EASY EXAMINATION QUESTIONS *[aside]*: We've done that already, on the quiet.

THE SHADE OF NEWTON *[obviously very interested]*: But my Second Law . . .

*(The announcement of the motion is greeted with mixed feelings.)*

CALCULUS: It's foolish. There's no point in differentiating between limits like that.

IMPULSE: You generally do have some axes to grind. But I must say the idea does bring one up with a jerk.

WORK: Why make things too easy? Hard work never killed anyone!

PURE GEOMETRY: I don't follow this line. The point I wish to make . . .

THE SHADE OF EUCLID *[as though wishing to remind a world which ignores his finer expositions]*: There isn't really a point at all.

HOT AIR *[rising]*: It's a sort of mirage, you know. But *[continuing, waveringly]* if we should happen to make a change round about now, it might possibly have something to do with causing a change in Einstein's Theory, if you know what I mean.

PHYSICS: Oh, don't bother to pay any attention to him, Hot Air always rises.

*(This is evidently the case. Most people have ignored Hot Air, except the Shade of Newton, who fainted at the mention of Einstein's Theory, and is now being revived by shouts of "Hoyle to thee, blithe spirit!")*

FORCE OF GRAVITY: We will put the matter to the vote . . .

CONIC SECTION *[projecting himself into the fray]*: About time we brought the matter to a focus.

FORCE OF GRAVITY: . . . and perhaps the Shade of Geiger will act as teller.

REAL VARIABLE: Isn't that one of the functions of $x$?

CIRCLE: Don't worry, it's as easy as $\pi$!

THE SHADE OF NAPIER: Yes it's as e'sy as falling off a log!

HYPERBOLIC FUNCTIONS: Cosh, it's a real sinh [*pronounced sinch*]!
FORCE OF GRAVITY: All in favour say "Aye".
COMPLEX VARIABLE: i.
CURRENT AND MOMENT OF INERTIA: *I*.
OPTICS: Eye.
CONDUCTOR: I offer no resistance . . . well [*catching the penetrating stare of X-RAY*] negligible, anyway.
MOMENTUM: I support a more conservative policy.
ENERGY: So do I on the whole . . .
AIR RESISTANCE *[aside, voicing the feelings of a small group of assorted Forces]*: Except when dissipated, the old so-and-so.
ENERGY: . . . but of course there are potential advantages.
SMALL INCREMENT: It's rather a great change to make now.
THE SHADE OF EINSTEIN: Only relatively speaking, of course.
COUPLE: I agree, this is not the right moment.
CHEMISTRY: It doesn't do to be too precipitate.
ELASTICITY: There's too much stress laid on $g$ — it is going beyond the limit.
FORCE OF GRAVITY *[catching the eye of REFLECTION]*: Well?
REFLECTION: I'm always the same as the person before me.

*(There are evidently more people against the motion than in favour, so it is abandoned. Conversation is resumed, until HEAT speaks.)*

HEAT: The dividend!
FORCE OF GRAVITY: Be more specific, Heat.
HEAT: I want a dividend. Other companies pay a dividend. Why not us?

*(This idea is popular and Heat is loudly supported.)*

DIVISOR: Quotient isn't here.
SUM: What difference does that make? There is no need to add difficulties.
DIVISOR *[patiently]*: He is the treasurer. If there is no quotient, the dividend cannot be distributed.
STATISTICS: Isn't there a normal distribution?
ALGEBRA: There are a good many factors to be taken into account.
COORDINATE GEOMETRY *[carelessly dropping a perpendicular on to his foot]*: Oh, cusp!
CHEMISTRY: OH OH, that's an unexpected reaction.
SMOOTH PLANE: My reactions are always normal.
BREADTH: I still don't understand why there is no dividend.
LENGTH: You always were a bit thick.
ELECTRICAL RESISTANCE: I think it's time we all went ohm.
ELECTROMOTIVE FORCE: What a revolting pun!

*(There is an interruption. INFINITY rushes across the room on his way to MINUS INFINITY, closely followed by REAL VALUES OF X. This breaks up the meeting. Everyone moves towards the door, except STATICS.)*

DYNAMICS *[at the door]*: Aren't you coming?
STATICS *[carefully adjusting the position of his centre of gravity]*: Shortly, but after all this activity, I just feel like a moment of inertia.

**Anon., with substantial additions by A. R. Pargeter.**

*"Life, yes — but as for intelligent life, I have my doubts."*

**Contributed by Andy Begg.**

## Hints for Pargeter's triangle (p. 43)

Put the numbers in prime factors. Look for horizontal connections rather than vertical ones.

For a further hint see p. 75.

## If negative rain falls on negative area . . .

Area: 41 100 km$^2$ (plus/minus 4.4 million hectares).
Rainfall: Highest during summer . . . with average 800 mm (plus/minus 32 inches) a year.

**From an advertisement for the Republic of Transkei in *The Times*, 26 October 1978. [Per B. Cook.] *MG*, 1979, p. 258.**

## Is it dying out?

It must also be noted that while death is less common today than it was a hundred years ago, it is increasingly confined to the elderly . . .

***Open Humanist*, August 1980.**

The group was alarmed to find that if you are a labourer, cleaner or dock worker you are twice as likely to die than a member of the professional classes.

***The Sunday Times*, 31 August 1980.**

People were dying less often than they were before.

**From a schoolgirl's essay on the Industrial Revolution. [All three per Colin Mills.] *MG*, 1981, p. 86.**

## Beefed up figures

All beef Beef Burgers (containing 90% beef).

**From a packet of Findus Beef Burgers. [Per P. H. Ransom.] *MG*, 1983, p. 196.**

## It pays to stay single

Single Membership is now $9.50 per annum. To facilitate 'families' of two we have introduced Couples Membership at a cost of $19.50.

**From an advertisement for the Hibernian Society (pointed out by Bruce Henry). *MG*, 1983, p. 200.**

## Sickly arithmetic

4% of nothing is nothing. We want 12%.

**From a poster during the health workers' dispute as reported in *The Daily Telegraph*, 15 June 1982. [Per A. R. Pargeter.] *MG*, 1984, p. 12.**

## Proof by induction?

The first Piccadilly Line train from Heathrow Central will not run on Sundays.

**From a notice at Manor House underground station (as reported by Dr Ian Livingstone). *MG*, 1983, p. 230.**

# Questions of logic

[Parts of questions from examinations in logic.]

Use the symbolism of the predicate calculus to indicate the logical basis of the following jokes:

(1) *Question*: Are there any congruent triangles in the following diagram?

*Answer*: Triangle ABC is congruent, but triangle DEF is not.

(2) A man staying at a hotel complained to the manager, "I don't like all those mice in my room". The manager replied, "Tell me which ones you do like and I will get rid of the rest".

(3) *Student*: I did not understand a word you said in that lecture.

*Lecturer*: Which word was that?

(4) *Question*: I have two coins and their total value is 6p. One of the coins is not a 5p. What are the coins?

*Answer*: Although one of the coins is not a 5p the other is a 5p. The two coins are 1p and 5p.

**Set by Keith Austin.**

# Back to the Eleven Plus

**Do you remember the sequence, said to have appeared in an 11+ examination: O T T F F S S E...? It has a brother (cousin?): 3 3 5 4 4 3 5 5...**

**A. R. Pargeter.**

# Nought for your comfort

Rather a natty sequence which has more to it than meets the eye, is 0 0 1 0 2 1 3 0 4 2 5 1 6 3 7 0 8 4...(No calculation needed — one could just as well use letters — but as given there *is* a simple formula!)

**A. R. Pargeter.**

## Quotelets from Bibby (*Quotes, Damned Quotes, . . .*)

If a coin falls heads repeatedly one hundred times; then the statistically ignorant would claim that the 'law of averages' must almost compel it to fall tails next time. Any statistician would point out the independence of each trial, and the uncertainty of the next outcome. But any fool can see that the coin must be double headed.

**Ludwik Drazek (1982).**

Arthur looked up. "Ford", he said, "there's an infinite number of monkeys outside who want to talk to us about this script for Hamlet they've worked out".

**Douglas Adams, *The Hitch-hikers Guide to the Galaxy.***

Man must learn to simplify, but not to the point of falsification.

**Aldous Huxley, *Brave New World.***

To guess is cheap, to guess wrongly is expensive.

**Old Chinese proverb.**

If all the statisticians in the world were laid head to toe, they wouldn't be able to reach a conclusion.

**Anon., after a comment on economists by G. B. Shaw.**

Facts speak louder than statistics.

**Mr Justice Streatfield (1950).**

He divided people into statisticians, people who knew about statistics, and people who didn't. He liked the middle group best. He didn't like the real statisticians much because they argued with him, and he thought people who didn't know any statistics were just animal life.

**Nigel Balchin, *The Small Back Room*, Collins, 1943.**

If there are three statisticians on a committee, there will be four minority reports.

**Anon.**

Computer warning: Always switch on brain before opening mouth.

**Anon.**

## Why are we here?

Mathematics may be defined as the subject in which we never know what we are talking about, nor whether what we are saying is true.

**Bertrand Russell, contributed by D. Brown, York and G. Tyson, London.**

Special Advertisement

# APPLIED MATHEMATICS KITS

As a result of recent reconstruction, the School Mathematics Project has a large amount of material for disposal at reduced prices, and offers to interested readers the following

SUPERB SURPLUS KITS IN
APPLIED MATHEMATICS

STANDARD KIT
includes

1 uniform ladder, complete with rough ground and smooth wall
1 hemispherical bowl of radius $a$, supplied together with
1 uniform rod, of length $2a \sin \theta$, with perfectly rough ends
2 particles, connected by a light inextensible string
2 rigid, smooth, incompressible, and perfectly elastic spheres, of masses $m$ and $em$ (masses $M$ and $eM$ can be supplied if preferred).
1 hank uniformly flexible string
1 set assorted frictionless massless pulleys, in uniform box, guaranteed not to influence acceleration.

SPECIAL "DE LUXE" KIT
includes, in addition,

1 rigid body, complete with axis, of moment of inertia $I$ (moments of inertia $A$, $B$, and $C$ can be supplied at a slight extra charge).
1 isosceles wedge with different coefficients of friction on all three faces, complete with pulley at apex (a connoisseur's item)
1 bottle of resisting medium, capable of extending to approximately $550gt/2240k$ feet
$n + 1$ bricks, ready stacked in a pile with uniform overlap.

---

The whole ready packed in a perfectly rough hollow sphere of radius $r$ and negligible thickness. Price $a + bx$, where $a$ and $b$ are constants and $x$ is the distance from the orchard at Woolsthorpe, Lincs.

Sherborne School — H. M. CUNDY *et al.*

**Advertisement in *MG*, 1963, p. 132.**

## Approximations

An approximate answer to the right problem is worth a good deal more than an exact answer to an approximate problem.

**John Tukey, contributed by Stig Olsson.**

## Ancient and modern

While modern mathematicians use a function,
And young mathematicians use an iterative process,
Old mathematicians use their fingers!

**Contributed by F. Tapson, Devon.**

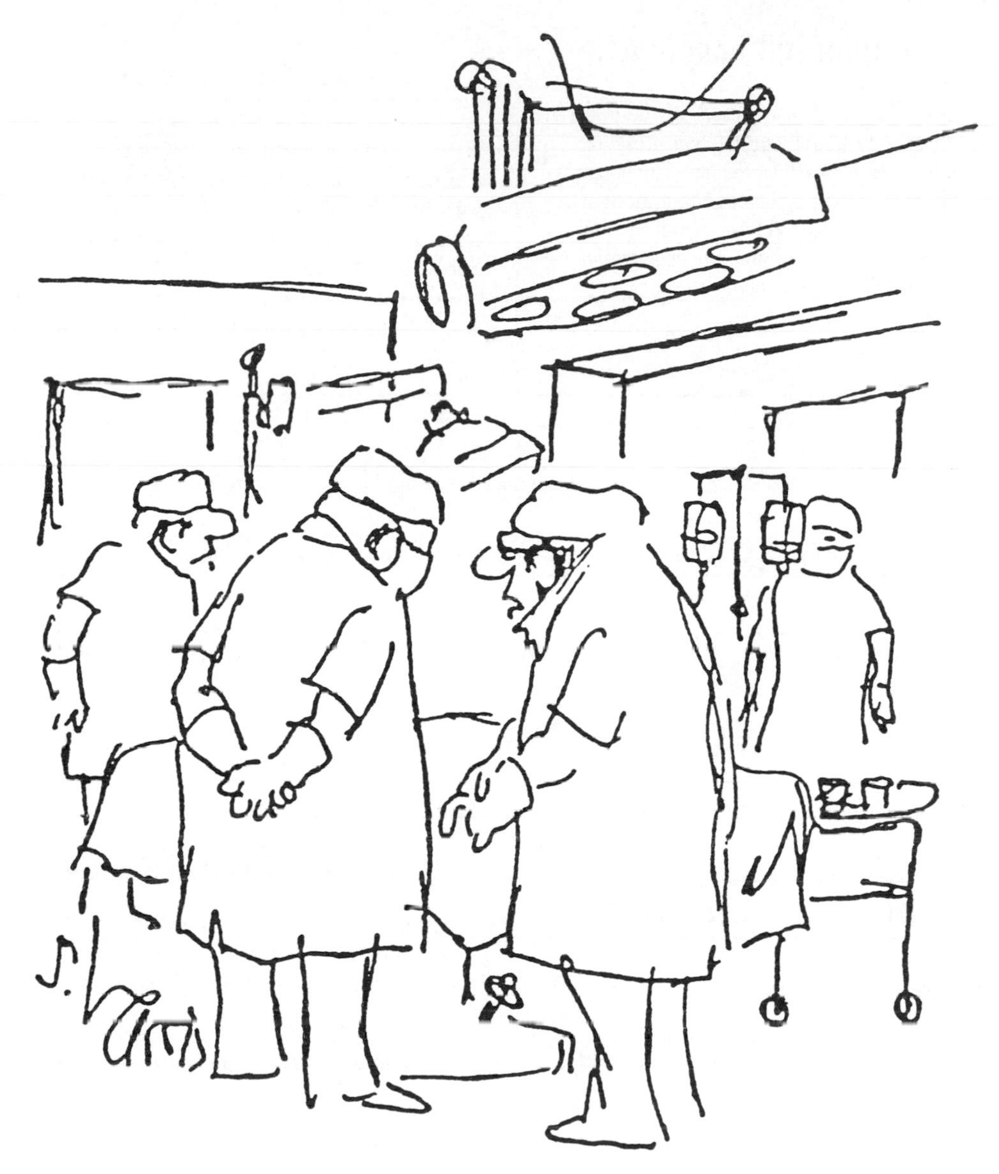

*"We'll only do 72% of it, since it's been reported that 28% of all surgery is unnecessary."*

**Contributed by Stig Olsson.**

## Overdone

To err is human, but when the eraser wears out ahead of the pencil, you're overdoing it.

**L. J. Peter, *Peter Prescription: how to be creative, confident and competent*, Bantam, 1973, contributed by Stig Olsson.**

## A standstill economy

You probably already know most of the ways of economising on petrol: correct tyre pressure; cruising at 60 rather than 70 mph; minimum use of choke; filling up before the motorway; buying lowest suitable octane; removing roof rack when not required; and not accelerating from standstill, which is about the most wasteful thing you can do.

***The Times*, 25 October 1980.**
**[Per David Blackman.] *MG*, 1981, p. 248.**

## Maths is Fun

Suppose that mathematics is not fun. The editors of this work would then have received no contributions, this book would not have been printed and you could not be reading it. You *are* reading this book. Therefore the initial assumption is contradicted, and mathematics is fun.

**An example of proof by *reductio ad absurdum*, contributed by D. Brown.**

## My favourite theorem by G. H. Hardy

My favourite theorem: If a parabola be a rectangular hyperbola then it is also an equiangular spiral. [Can be proved rather crudely by analysis, but the projective proof is neat.]

**Contributed by A. R. Pargeter.**

## Another even prime

The highest known prime number is $2^{21701}$.

**From *The Guinness Book of Records*, 1980.**
**[Per D. Delmar.] *MG*, 1981, p. 61.**

## Unattached free end

A conical pendulum consists of a light inextensible string of length $L$ with a particle of mass $M$ attached to its friend . . .

**Printing error in an exam paper. [Per D. Delmar.] *MG*, 1981, p. 202.**

## The rule of three?

Multiplication is vexation,
Division is as bad;
The rule of three doth puzzle me,
And practice drives me mad!

**[The rule of three and practice are old arithmetical methods.]**
**Trad., contributed by A. R. Pargeter.**

## Unexpected factors

Factorise $a^3 + 3a^2 + 3a + 2$ and $a^4 + a^3 + a^2 + 2$.

[$(a + 2)(a^2 + a + 1)$ and $(a^2 - a + 1)(a^2 + 2a + 2)$]

**Contributed by A. R. Pargeter.**

# Queens of arithmetic

"Can you do multiplication?" asked the Red Queen. "What is 3 multiplied by 2?"
"That's 6," said Alice.
"Wrong," said the Red Queen: "it's 11."
"But that's not twice three," said Alice.
"Never mind, dear," said the White Queen. "We have better multiplication than you, to multiply 3 by 2 we say: twice three is 6, 2 and 3 make 5, and then 6 and 5 make 11. Now you try one. Do 11 multiplied by 4."
"Well, four elevens are 44," said Alice: "11 and 4 make 15, and 44 and 15 make 59. Is it 59?"
"Right, for once," said the Red Queen.
"But multiplication doesn't mean all that," said Alice.
"We say it does," said the Red Queen. "Humpty Dumpty pays the word well and it does its duty."
Alice remembered some of her Modern Mathematics lessons. "We have some Laws about Multiplication. Commutative and Associative, I think they are called."
"So have we," said the Red Queen. "Try 3 multiplied by 2 and 2 multiplied by 3. It comes to 11 either way, doesn't it?"
"Yes," said Alice, rather dispiritedly.
"Try 2 multiplied by 3 multiplied by 4. 2 multiplied by 3 is 11, 11 multiplied by 4 is 59. Now try it the other way. 3 multiplied by 4 is 19, 2 multiplied by 19 is 59. There you are."
"I suppose so," said Alice and lapsed into silence.
But she soon brightened up. "We have another law," she said: "the Distributive Law. Now I don't believe that Addition is distributive over your Multiplication."
The Red Queen snorted. "I should think not, with your Addition. But with ours it is. Add one and one."
"Two," said Alice.
"Wrong again. It is nought," said the Red Queen.
"One and one is rather difficult," said the White Queen kindly. "Try something else. Add 29 and 44."
"What is it?" asked Alice. "How can I tell?"
"It is 17," said the Red Queen: "and you can tell by looking at Humpty Dumpty's tables." She handed a paper to Alice and then, turning to the White Queen, said "We shall know soon." The White Queen got up, walked off along the path and disappeared through the gate.
Alice looked at the paper. It was headed "Odds and Ends, by Humpty Dumpty," and there followed such things as

$$\begin{aligned} 1 +_0 2 &= \tfrac{1}{5}, \\ 2 +_0 2 &= \tfrac{1}{2}, \\ 2 +_0 5 &= 1, \\ 3 +_0 3 &= 1, \\ 3 +_0 4 &= 1\tfrac{2}{9}, \\ 3 +_0 11 &= 2, \\ 5 +_0 11 &= 3, \text{ and so on.} \end{aligned}$$

Underneath was $m +_0 n = \dfrac{mn - 1}{m + n + 2}$

Alice did not know what to make of this. She handed back the paper and said "I like our Arithmetic better. This seems very odd."

"Odds," corrected the Red Queen.

"Where are the ends?" asked Alice.

"Right at the bottom. What do you see there?" said the Red Queen.

"Why, nothing," said Alice.

"Ah," said the Red Queen: "You want to know the other End, what it is for. Now look at this." She handed Alice another piece of paper.

Alice read: "Humpty Dumpty's Odds; Racing of the King's Horses." Then there were some lists of names and numbers. Alice noticed some underlined items. They were:

2.30 Mutton Pies 9:1, Jam To-Morrow 14:1.
3.00 White Knight 2:1.

The Red Queen spoke with unusual animation. "We've backed *either* Mutton Pies *or* Jam To-Morrow to win the 2.30 and White Knight to win the 3.0. Now we can find the odds against this compound event. 9 oddly-added to 14 makes 5; then 5 oddly-multiplied by 2 comes to 17. So it is 17 to 1 against. Or we can say 9 oddly-multiplied by 2 is 29, 14 oddly-multiplied by 2 is 44, and 29 oddly-added to 44 is 17. As we have staked sixpence [$2\frac{1}{2}$p] we may win 8/6 [$42\frac{1}{2}$p]."

Alice tried once more. "Is your odd-addition commutative and associative? Are there unit elements?"

The Red Queen would not respond. "We've said enough about all that. Now if you have some money why don't you —".

But a heavy crash shook the forest from end to end and she never finished the sentence. They waited apprehensively in silence for a moment, and then they saw the White Queen hurrying back.

"It's no good," she panted: "He's broke."

**R. H. Cobb, Malvern College, Worcs., *MG*, 1963, pp. 128–30.**

# Pen-sketch

Professor Surd, the well-known mathematician, lives at the intersection of the crossroads in a log cabin named "Mantissa" (presumably because it is without characteristic) together with his family of six children. "The smallest perfect number," he said when I visited him the other day, "and two of them twin primes". Through a lattice window they could be seen in a group, safely contained in a near field. He had just finished locating the roots of an old plane-tree (for which he intended to substitute a ring of roses), in company with his faithful hound Vector who, from his origin, had been trained to carry the Professor's displaced constructs, and appeared to have a good sense of direction.

Projecting his old hemispherical cap on to a smooth peg, he removed the convex cover from a rigid framework which turned out to be a chair, and offered me the seat with the words "Park your hull". A basis for the ensuing discussion was provided by the remarkably irrotational motion of the inviscid beverage served to us by Mrs Hernia Surd. "Made to my own formula," said the Professor, in reply to my direct complement, "from primitive elements grown in my own field". Does he hope to inject his discovery into the tea-drinking set with any chance of success? "Pi in the sky," he said — an oblique remark, for we all know that he seldom descends to figures of speech.

**Contributed by F. Gerrish, Woking.**

# Questions for a political science recruit

Are you sure you're up to date?
Do you explain, or explicate?
Do you guess, or hypothesize,
And can you content-analyze?
Do you repeat, or replicate,
Do your parameters obfuscate?
Do you work with facts or data,
And have your variables an indicator?
Have you dropped sight for perception,
Is your every notion a conception?
Do you look for proof, or validation,
Can you manage disconfirmation?
Has societal superseded social
And affective banished the emotional,
Have all imprecisions really gone
From your scientific lexicon?
Is your focus sub-systemic?
Have you transcended mere polemic,
Eschewed intuition and impression,
And made of science your profession?
Can you achieve a minor miracle
By being rigorous and empirical?
Come now, can you operationalize,
Quantify and conceptualize?
Can your output be machine-read?
Have you a code in your head?
Are you adept at research design —
Brother can you paradigm?

**I. L. Claude Jr., quoted in *Quotes, Damned Quotes, . . . ,* ed. John Bibby.**

## Biblical fingers get stuck into pi

Noting that there are two references to the mathematical constant pi (= 3.14159 . . .) in the Bible (I Kings 7:23 and II Chronicles 4:2), both of which imply the value pi = 3, some academics in Kansas have started the Institute for Pi Research, whose avowed intention is to propagate the usage of the value of 3 for pi. As the Institute's founder, Samuel Dicks, Professor of Medieval History at Emporia State University says: "If a pi of 3 is good enough for the Bible, it is good enough for modern man."

The Institute is campaigning for the value of pi = 3 to be given equal time with the more conventional value in state schools. Coupled with Dick's statement that: "I think we deserve to be taken as seriously as Creationists," this would appear to give some clue as to what really lies behind the foundation of the Institute. As economic historian Loren Pennington says, "If the Bible is right in biology, it's right in math."

But what ever their real aim, they seem to have friends in high places. The group wrote to President Reagan asking for his support, and though they did not receive a reply, they were greatly encouraged to hear him say in a speech shortly afterwards that "The pi(e) isn't as big as we think."

***The Guardian*, 21 June 1984, reproduced by permission. Contributed Keith Selkirk.**

*"It says it has discovered a very nice proof but its memory is insufficient to hold it."*

***Mathematical Magazine*, Mathematical Association of America.**

## Have we missed some integers?

Complete the sequences:

$\sqrt{4}, \sqrt{3}, \sqrt{2}, x, \sqrt{1}$; and $\sqrt{1}, \sqrt{2}, \sqrt{3}, y, \sqrt{4}$.

**[The solutions are 2 sin ($\pi/5$) and 2 sin ($5\pi/12$).]**
**Contributed by A. R. Pargeter.**

## Geometric tour

Square–Square (Circular route).

**From a summary of Bournemouth's bus services.**
**[Per D. Delmar.] *MG*, 1981, p. 166.**

## Multiplication before logs

Before the invention of logarithms, two methods to simplify multiplication were available:

(a) a table of quarter squares using the formula:

$$ab = \tfrac{1}{4}(a + b)^2 - \tfrac{1}{4}(a - b)^2$$

(b) a table of cosines, using the formula:

$$\cos A \cos B = \tfrac{1}{2}\{\cos(A + B) + \cos(A - B)\}$$

**Contributed by A. R. Pargeter.**

## Budget

University President: 'Why is it that you physicists always require so much expensive equipment? Now the Department of Mathematics requires nothing but money for paper, pencils, and erasers . . . and the Department of Philosophy is better still. It doesn't even ask for erasers.'

**From Isaac Asimov, *Treasury of Humour*, Woburn Press, London, 1971, quoted in *More Random Walks in Science*, ed. R. L. Weber, Institute of Physics, Bristol, 1982.**

# Ask a silly question

***Punch.***

# The discovery of the mathematics textbook

*February, 2003 a wet Friday afternoon:*
Form 2B are seated at their individual computers, tapping keys, running their mathematics instruction programmes. They look bored. There have been so many lessons just like this one with nothing to do except press keys and watch a display appear on the screen. The class teacher, Mr Conic, is turning out a cupboard. He is nearing retirement (30 next birthday — compulsory retirement for all teachers was introduced in 1996) and is tidying up prior to his departure. Among the tapes, disks and broken computers he comes across an old mathematics textbook published some 35 years ago. As he browses through the book, John, a bright pupil, comes out to tell him that his computer has just blown up and could he please have another one. John notices the book and asks if he can borrow it for the weekend.

*The following Monday:*
As 2B come in reluctantly to sit at their machines, John rushes excitedly to Mr Conic. "Sir, can I use the book instead of my computer? It's good." Mr Conic hesitantly agrees and a search ensues to find paper and pencil when it is realised that John will not be able to work with his keyboard and screen. The other children are inquisitive and a little bit jealous.

*Tuesday:*
Mr Conic is greeted by a chorus of requests, "Please Sir, can we have a go with the book?" In the face of mounting pupil pressure he decides to allow two children at a time to share the book and the paper and pencil. The children have never before shown such interest and excitement.

*2 weeks later:*
The headteacher appears in Mr Conic's room. He has received numerous complaints from parents about their children wasting time with a book in their mathematics lessons. But he comes away impressed by the children's enthusiasm for the textbook and indeed even promises to ask the PTA for money to buy more pencils.

*1 month later:*
The news of Mr Conic's bold, innovative teaching has now travelled far. The local adviser has been in. Other schools in the area have tried to dig out old texts and to follow Mr Conic's lead. Today a team of HMIs visits the school, they depart impressed. Mr Conic is asked to consider remaining beyond the normal retiring age and to lecture up and down the country on the use of the textbook.

*6 months later:*
The DES announces that it would like to see a textbook in every school and that it will pay half the cost of a book if the school will raise the money for the other half. Research projects, to monitor the use of textbooks, are announced at two of the remaining three universities in the United Kingdom.

*One year later:*
With only one book in a classroom, teachers are finding problems in organising pupils. It is hard for more than two children at a time to use the book. In classrooms all over the country pupils eagerly await their turn to use a book.

Special in-service courses for teachers are established. Some schools claim success in using whole sets of books with all pupils working at the same topic at the same time. Children observed *discussing* their work and abandoning their machines. Research findings are inconclusive. Professor Conic returns from a highly successful American lecture tour.

*5 years later:*
Most schools are now using textbooks. Cupboards are full of old computers. On a wet Friday afternoon in February. Form 2B sit with bored looks on their faces as they struggle through monotonous exercises in their books. The only prospect of relief is to obtain permission to sharpen a pencil.

A major DES research project on the use of textbooks in the teaching of mathematics has just failed to reach any definite conclusions. Concern about the state of school mathematics has led to the setting up of a Committee of Enquiry into the teaching of mathematics under the chairmanship of Dr Croftcock — Professor Conic is to be a member of the committee.

*10 years later:*
Mr Conic, son of Professor Conic, is teaching 2B. The pupils are all working wearily at their textbooks. Mr Conic is turning out an old cupboard. Just as he unearths an old computer, John, a bright pupil, comes out to tell him that his textbook has disintegrated . . .

**Michael Cornelius, Durham, in *Mathematics in School*, September 1983.**

**Vic Norton, University of Miami.**

## Slicker than most (or a new version of the oil drop experiment)

A 10-mile slick off the Japanese coast was caused by three and a half gallons of oil from a British bulk carrier . . . magistrates at Uxbridge, Middlesex, were told yesterday.

**_The Daily Telegraph_, 28th June 1980. [Per Frank Budden.] _MG_, 1981, p. 252.**

## A sporting chance

This winning the toss business in overs cricket is a very dicey business.

**Said before the Surrey v. Yorkshire semifinal of the 1980 Gillette Cup by Don Brennan (as reported by Colin Dixon). _MG_, 1982, p. 198.**

## Including flow charts?

Baths up to A level by experienced successful private tutor.

**Small ad. in _The Times_, 20 October 1981, spotted by G. H. Bailey. _MG_, 1982, p. 250.**

## Mathematical entertainment

Because mathematicians get along with common words, many amusing ambiguities arise. For instance, the word *function* probably expresses the most important idea in the whole history of mathematics. Yet, most people hearing it would think of a 'function' as meaning an evening social affair, while others, less socially minded, would think of their livers.

**E. Kasner and J. Newman, _Mathematics and the Imagination_, Bell, London, 1968, quoted in _More Random Walks in Science_, ed. R. L. Weber, Institute of Physics, Bristol, 1982.**

## Justifying educational research

Understanding atomic physics is child's play compared to understanding child's play.

**Anon., quoted in _More Random Walks in Science_, ed. R. L. Weber, Institute of Physics, Bristol, 1982.**

# Come and teach maths – if you've the head for it

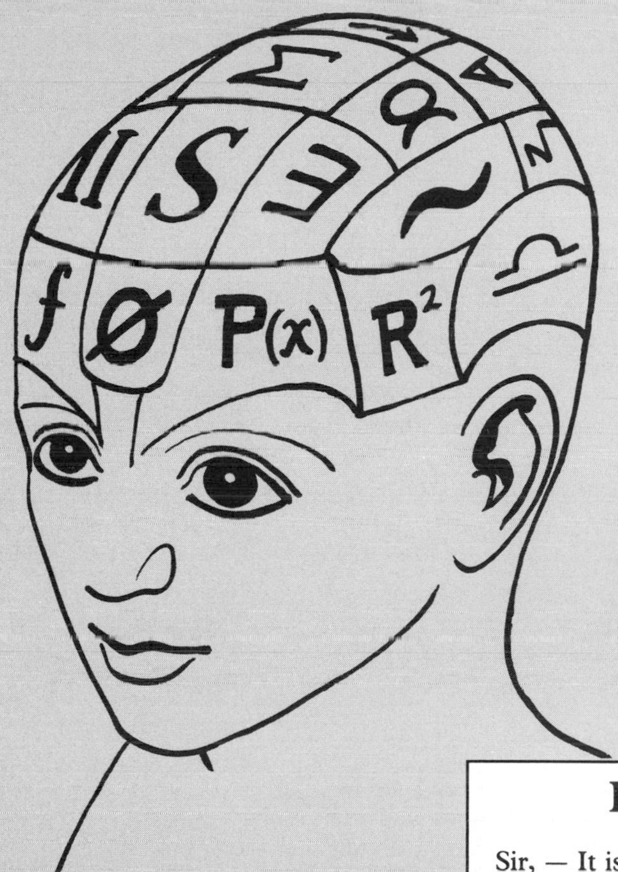


The future is good for mathem
education is recognised as a vi
development, the progress on
depend. Taken as a whole—th
freedom of choice of place of w
the satisfaction of doing a wort
favourably with any job open t

Over 40% of maths and scie
seven years ago are already h
over 40 is a head teacher, while
good use in administration or i
are available in almost every p
holidays.

**Pay and prospects.** Startin
other careers, ranging for grad
forward, as head of a large de
deputy headship could take
like responsibility, is seldom
teacher. And if one is lookin
at the educational service as a
experience is needed are in the
teaching? Read our free bookl
Available from: The Departm
T.E.S.3, Room 107), Curzon St

## Empty heads

Sir, — It is unfortunate that whoever designed the "head for maths" in the current advertisement of the D.E.S., should have placed in the centre of the forehead the accepted symbol (∅) for the empty set!

A. R. Pargeter, Tiverton, Devon.

# The kiss precise

If $a$, $b$, $c$ and $d$ are the reciprocals of the radii of four circles in a plane, each of which touches the other three, then

$$2(a^2 + b^2 + c^2 + d^2) = (a + b + c + d)^2$$

For pairs of lips to kiss maybe
Involves no trigonometry.
'Tis not so when four circles kiss
Each one the other three.
To bring this off the four must be
As three in one or one in three
If one in three, beyond a doubt
Each gets three kisses from without.
If three in one, then is that one
Thrice kissed internally.

Four circles to the kissing come.
The smaller are the benter.
The bend is just the inverse of
The distance from the centre.
Though their intrigue left Euclid dumb
There's now no need for rule of thumb.
Since zero bend's a dead straight line
And concave bends have minus sign,
*The sum of the squares of all four bends*
*Is half the square of their sum.*

To spy out spherical affairs
An oscular surveyor
Might find the task laborious,
The sphere is much the gayer,
And now besides the pair of pairs
A fifth sphere in the kissing shares.
Yet, signs and zero as before,
For each to kiss the other four
*The square of the sum of all five bends*
*Is thrice the sum of their squares.*

And let us not confine our cares
To simple circles, planes and spheres,
But rise to hyper flats and bends
Where kissing multiple appears.
In $n$-ic space the kissing pairs
Are hyper spheres, and Truth declares —
As $n + 2$ such osculate
Each with an $(n + 1)$-fold mate.
*The square of the sum of all the bends*
*Is* n *times the sum of their squares.*

**Frederick Soddy in *Nature*, 20 June 1936, last verse by Thorold Gosset in *Nature*, 9 January 1937, and reproduced by permission.**

## The hexlet (a sequel)

However ill-assorted in girth three spheres may be
Each one can kiss the other two simultaneously
A ring of six about them all kissing serially.

Though any necklet of graded beads
May fit in general the she-sex,
This hexlet of mine of novel design
Caresses not one but three necks.
However it's worn it alters its grade
To suit its tri-spherical prison,
Plays kiss-in-the-ring and merry-go-round
Whilst hugging three necks with precision.
Like bubbles that blow and kindle and go
It holds up to light-hearted derision
Makes of the pure circumflex
And its pet aversion in the mental inversion
That will have "It's $1/x$."

All saints and sages throughout the ages
From one doxy never have swerved,
To hold fast unto what in change changes not
And ferret out what is conserved.
Now these beads without flaw obey this first law
For the aggregate sum of their bends.
As each in the tunnel slims through the funnel
Its *vis-à-vis* grossly distends.
*But the mean of the bends of each opposite pair*
*Is the sum of the three of the thoroughfare.*

Frederick Soddy in *Nature*, 5 December 1936, and reproduced by permission.

## Three in one . . .

The Holy Trinity is like the cube roots of unity — one real, two imaginary.

Anon., contributed by G. Tyson.

## Hint for Pargeter's triangle (p. 43)

Answers in lines 3 and 5 are both 7, in line 6 the answer is 227, and in line 7 it is 127.

## The plain truth

Truth is ever to be found in simplicity, and not in the multiplicity and confusion of things . . . He is the god of order and not of confusion.

**Sir Isaac Newton, quoted in *More Random Walks in Science*, ed. R. L. Weber, Institute of Physics, Bristol, 1982.**

## The bookworm

The three volumes of a book stand on a shelf in order. Each book has a set of pages one inch think and two covers each $\frac{1}{8}$ inch thick. A bookworm eats its way from the first page of volume one to the last page of volume three, following as short a line as possible. How far does it go.

[Answer: $1\frac{1}{2}$ inches].

**Contributed by A. R. Pargeter, modified by Keith Selkirk.**

## The Ten Commandments of Statistical Inference

I. Thou shalt not hunt statistical inference with a shotgun.

II. Thou shalt not enter the valley of the methods of inference without an experimental design.

III. Thou shalt not make statistical inference in the absence of a model.

IV. Thou shalt honour the assumptions of thy model.

V. Thou shalt not adulterate thy model to obtain significant results.

VI. Thou shalt not covet thy colleagues' data.

VII. Thou shalt not bear false witness against thy control group.

VIII. Thou shalt not worship the 0.05 significance level.

IX. Thou shalt not apply large sample approximations in vain.

X. Thou shalt not infer causal relationships from statistical significance.

**Michael F. Driscoll, in *The American Mathematical Monthly*, 84:628 (1977), quoted in John Bibby, *Quotes, Damned Quotes and . . .***

## Small is beautiful

The purpose of the visit was . . . to observe the work of a small sample student and the contribution he or she made to the course.

**From the report of the research project on the Mathematical Association Diploma. [Per Douglas Quadling.] *MG*, 1983, p. 241.**

*THE TETRAHEDRON — a new member of the percussion family which brings an extra dimension into musical interpretation.*

(With apologies to Gerard Hoffnung.)
*Mathematical Digest*, University of Cape Town.

*"Look — if you have five pocket calculators and I take two away, how many have you got left?"*

## Trig is easy!

$$\sin^2 5\theta - \sin^2 3\theta = (\sin 5\theta + \sin 3\theta)(\sin 5\theta - \sin 3\theta)$$
$$= \sin 8\theta \sin 2\theta$$

[Note: the result is correct and works for any numbers, but deeper justification is needed.]

**Contributed by A. R. Pargeter.**

## A variable constant

$$\frac{\mathrm{d}}{\mathrm{d}x}(\exp x) = \mathrm{ep}\,x + \exp.$$

**Contributed by Michael Mudge, Wolverhampton.**

## i is real

Since $\cos x = \frac{1}{2}(e^{ix} + e^{-ix})$ and $i \sin x = \frac{1}{2}(e^{ix} - e^{-ix})$, and also $\cosh x = \frac{1}{2}(e^{x} + e^{-x})$ and $\sinh x = \frac{1}{2}(e^{x} - e^{-x})$, we obtain $i \sinh x = \sin ix$ and $\cosh x = \cos ix$. From the second of these we obtain: $i = \{\cos^{-1}(\cosh x)\}/x$.

[Note: consider the values of $x$ for which the relations are defined.]

**Contributed by J. R. Harris, Isle of Wight.**

## Pre-war money exchange

Show that, in terms of pre-decimal currency,

£$i^i \approx$ 4s. 2d. [just under 21 pence]

where $i$ = the square root of $-1$.

[This was the old rate of dollar exchange when the UK and the US saw i to i.]

**Contributed by Frank Chorlton, Aston University.**

## Another way of obtaining (almost) the same answer

£$i^i$ may look rather imaginary, but is just over 20p. If $z = i^i$, taking principal values of logarithms, $\ln z = i \ln i = i(i.\frac{1}{2}\pi)$. Hence $z = e^{-\pi/2} = 0.208$ (to three decimal places). This is 4s. 1.9d.

**Contributed by Valerie de la Praudière, after Brian Bolt.**

## Is there a point?

What are the next three terms of the sequence:

Aggie, Aggie, Aggie, Betty, Betty, Betty, Cath, Cath, Cath, ., ., .?

[Answer: Dot, Dot, Dot.]

**Contributed by Keith Austin.**

## One, two, three blind mice

In some places there just aren't enough dormice. In other places they seem to be reasonably numerate.

**From Radio 4's *Natural History Programme* as reported by P. C. Goodwin. *MG*, 1987, p. 292.**

## The perfect compass

If one could make a perfectly smooth, perfectly round, non-magnetic sphere, it would be impossible to rotate it. If it also had a transparent surface with markings underneath, it could be used as a compass, provided one did not take it too far from its place of manufacture.

**Contributed by A. R. Pargeter.**

## Oops!

Write down in figures eleven thousand, eleven hundred and eleven.

**Contributed by A. R. Pargeter.**

## Sums and products

If $24n - 1 = ab$, and $a$, $b$ and $n$ are all integers, then $a + b$ is a multiple of 24.

**Contributed by A. R. Pargeter.**

## Polish serenade

A plane took off from Warsaw with its full quota of passengers. After a few hours, both pilots went down with food poisoning — so the chief stewardess rushed into the passenger section and asked "Does anyone know how to fly a plane?" A passenger stood up and went with her into the cockpit. After a while looking at all the dials and switches, he said "Well, we're going to crash." "What!" came the reply, "I thought you could fly a plane." "No," said the man, "I'm only a simple Pole in a complex plane."

**Contributed by E. S. Ratley, Essex.**

## I do too

Sir — Following Mr C. D. Tracey's letter about strange numbers, how about dividing the first nine digits (reversed) by the first nine digits (unreversed)? 987654321 divided by 123456789 is exactly 8. I find this curious.

**From a letter to *The Daily Telegraph*, 30 October 1985, spotted by A. W. Ingleton. *MG*, 1986, p. 42.**

## I missed that answer

If $2^x = x^2$, then $x = 2, 4$ or $-0.766\,664\,7\ldots$

Contributed by A. R. Pargeter.

Contributed by Michael Mudge, Wolverhampton.

## Caterwauling

Two cats are sitting on a sloping roof. Which is the more likely to slip off?

[Answer: The one with the smaller $\mu$.]

Contributed by A. R. Pargeter, Devon and M. R. Mudge, Wolverhampton.

## Maths rules O.K.

The logical sequel to "Maths rules O.K." is "Maths protractors O.K.", which is another statement about stocktaking in the mathematics department.

Contributed by J. R. Bradshaw, Tamworth.

## The new secs?

Around a quarter to one third of the women recruited to IBM and ICL as graduate trainees are female.

From *The Sunday Times*, 4 November 1984, sent in by Rosalie McCrossan. *MG*, 1985, p. 171.

## By definition

Even the biggest triangle has only three sides.

*The New Statesman*. [Per Dr A. K. Austin.] *MG*, 1969, p. 345.

## Why bother?

Mathematics contains much that will neither hurt one if one does not know it nor help one if one does know it.

J. B. Mencken, *De Charlataneria Eruditorum*, 1715. Quoted in *Fantasia Mathematica* by Clifton Fadiman, Simon & Schuster, New York, 1958.

## Very mean

The smallest integers with integral arithmetic, geometric and harmonic means are 10 and 40.

**Contributed by A. R. Pargeter.**

## Squares and Cubes

If the square of an integer is the difference between two successive cubes, then the integer itself is the sum of two successive squares.

**Contributed by A. R. Pargeter.**

## Pre-natal statistics

Dr J. Skone said 'A study early this year showed that nearly ten per cent of maternity patients in South Glamorgan were over 20 weeks pregnant'.

***The South Wales Echo*, 15 August 1979. [Per Alan Cohen.] *MG*, 1980, p. 259.**

## We're not surprised

The Yancos of the Amazon have for 3 the word POETTARRARORINCOAROAE. They do not count beyond 3.

**Source unknown.**

## Equality

All animals are equal, but some are more equal than others.

**George Orwell, *Animal Farm*.**

## With-it integers

The nearest integer to $n + \sqrt{n}$ is never a square.

**Contributed by A. R. Pargeter.**

## $d^3p/dt^3 > 0$

Rate of inflation accelerates.

**Headline in *The Financial Times*, 13 November 1973. [Per G. C. Shephard, who suggests that this may be the first time that news about a third derivative has made the front page of a national newspaper.] *MG*, 1976, p. 262.**

## Decimalisation of the Calendar

The American Mathematical Monthly . . . is published ten times a year.

**From an advertisement in *MG*, March 1979. [Per J. E. Drummond.] *MG*, 1980, p. 267.**

## That accounts for the backlog

On an average week-day 4.5 million letters and 100,000 parcels are posted and 3.3 million letters and 79,000 parcels are delivered.

**From a Post Office hand-out. [Per Gillian Hatch.] *MG*, 1984, p. 56.**

## You can say that again

Since only 6 centres submitted Statistics statistics, Statistics statistics statistically are unreliable.

**From a Southern Regional Board's report. [Per C. W. O. Rainer.] *MG*, 1983, p. 275.**

## At least we're good at something

In 'The Record Breakers' (BBC-1, 5.10) compere Roy Castle will be opening the Most Boring Book in the World — about mathematics.

**From a popular daily newspaper. [Per Peter Avery.] *MG*, 1976, p. 309.**

## Anti-clock-wise

The marvels of nature, quoth he,
Are ever a wonder to me.
That each tick and each tock,
Of a grandfather clock,
Is $2\pi$ root $l$ over $g$.

**Anon., contributed by John Backhouse, University of Oxford.**

## No polymaths?

Numbers worry polytechnics.

**A headline from *The Times*, 4 August 1983. [Per G. H. Bailey.] *MG*, 1984, p. 178.**

## A conservation problem

A mathematician and his wife (or, if you prefer it, her husband) entered the kitchen. The spouse pointed at an empty plate and said "Where has the fish gone?" They looked round and saw their cat on the floor. "How much fish was there?" "Three pounds." The mathematician picked up the cat and put it on the scales; they read three pounds. "We have found the fish. The only problem is where has the cat gone?"

**Contributed by Keith Austin.**

## The latest fashion?

". . . in this branch of the subject it is very difficult to be sure of water-proof tights."

**Almost said by Professor Goldstein during a lecture. Contributed by A. R. Pargeter.**

## Not Dr Who

I could be bounde in a nutshell, and count myself a king of infinite space.

***Hamlet*. Contributed by A. R. Pargeter.**

## A new axiom

"Parallel lines are lines which never meet if produced far enough."

**Anonymous pupil. Contributed by A. R. Pargeter.**

## Third?..., ninth?..., *n*th?

Our Scotland Yard correspondent writes . . . use of this equipment would change that tradition and we will hold back to the ninth degree before we do use it.

***The Daily Telegraph*, 16 July 1981. [Per S. Nelson Taylor.] *MG*, 1983, p. 13.**

## German inefficiency

Commenting on German efficiency Draper told how he had ordered $2\frac{1}{2}$ million Wotan flash cubes — which normally came packed in threes. To make up the precise quantity Wotan included one pack containing only two cubes.

**A report in *What Camera Weekly*, 26 December 1981 (spotted by M. Markey). *MG*, 1983, p. 139.**

*"Figure 1 is called a 'right angle' and you would naturally suppose that Figure 2 is a 'left angle', but according to the National Curriculum this is also a right angle."*

## Are computers replacing mental arithmetic?

But Mr Hilditch said computer research had shown that, given a 30 per cent discount in fares, an operator needed a 43 per cent increase in passengers to get back to the starting position.

***The Leicester Mercury*. [Per Eric Primrose.] *MG*, 1983, p. 122.**

## $T$ or $dT/dt$?

The time is rapidly approaching 8 o'clock.

**Heard on Radio 3 by J. G. Brennan, *MG*, 1983, p. 92.**

## Not from a normal family

In this new edition the late Professor Nevanlinna has added a number of sections on normal families.

**From a leaflet advertising the latest edition of *Introduction to Complex Analysis* by R. Nevanlinna and V. Paatero. [Per Prof. J. S. Pym.] *MG*, 1983, p. 188.**

# Einstein anticipated

On would think that besides infinit Space there could be no more Room for any Treasure. Yet to show that God is infinitly infinit, there is Infinit room besides, and perhaps a mor Wonderful Region making this to be infinitly Infinit. No man will believ that besides the Space from the Centre of the earth to the utmost bounds of the Everlasting Hills, there should be any more. Beyond those Bounds perhaps there may, but besides all that Space that is illimited and present before us, and absolutly endles evry Way, where can there be any room for more? This is the Space that is at this Moment only present before our Ey, the only Space that was, or that will be, from Everlasting to Everlasting. This Moment Exhibits infinit Space, but there is a Space also Wherin all Moments are infinitly Exhibited, and the Everlasting Duration of infinit Space is another Region and Room of Joys. Wherein all Ages appear together, all Occurences stand up at once, and the innumerable and Endless Myriads of yeers that were before the Creation, and will be after the World is ended are Objected as a Clear and Stable Object, whose several Parts extended out at length, giv an inward Infinity to this Moment, and compose an Eternity that is seen by all Comprehensors and Enjoyers.

**Thomas Traherne, *Centuries of Meditations*, The Fifth Century, paragraph 6, quoted in *Poems, Centuries, and Three Thanksgivings*, ed. Anne Ridler, Oxford University Press, 1966, pp. 369–70 and reproduced by permission. A quite astonishing anticipation of the notion of 4-dimensional space-time by a seventeenth century divine (the son of a shoemaker; born in 1637; went to Brasenose College, Oxford (B.A. 1656); died 1674.) Contributed by A. R. Pargeter.**

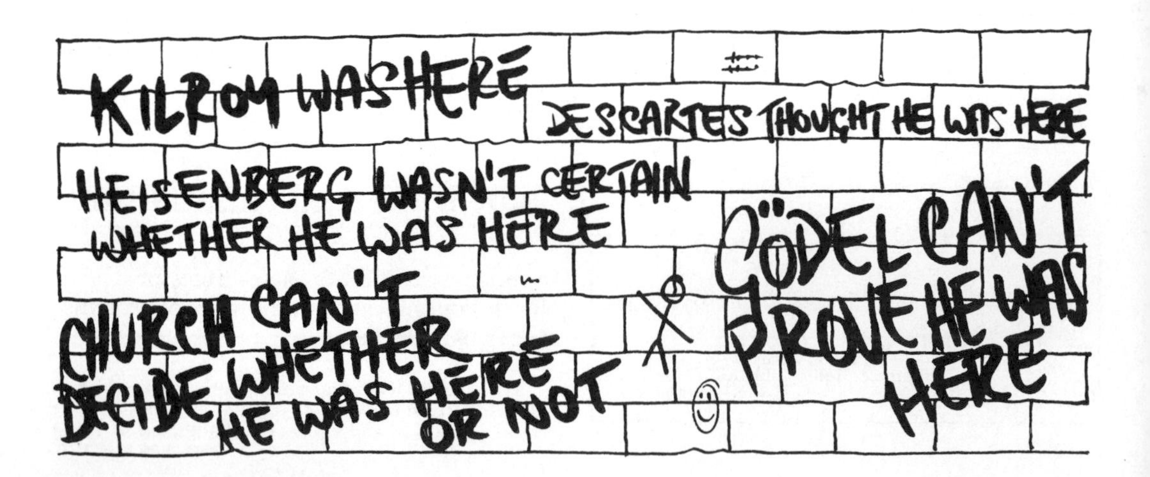

## CONTRIBUTORS

We apologise to any contributors who have inadvertently been omitted from the following list, and express our thanks to all of those who contributed, even though it was not possible to publish some contributions.

Keith Austin, Sheffield
John Backhouse, Oxford
Andy Begg, Wellington, New Zealand
John Bibby, Edinburgh
Neil Bibby, London
J. R. Bradshaw, Tamworth
D. Brown, York
Tom Bunting, Jersey, CI
David Chandler, Newbury
Frank Chorlton, Aston
Charles Cooke, Nottingham
Brian Cooper, Birmingham
M. L. Cooper, East Ham
B. Ll. Cutler, Warwick
John Deft, Bristol
Tony Gardiner, Birmingham
F. Gerrish, Woking
Simon Gray
E. Gulbenkian, Surrey
J. R. Harris, Isle of Wight
Sidney Harris, New Haven, USA
John Hersee, Bristol
Peter Jones, Nottingham
J. P. Knee, Malvern
Matthew Linton, Hong Kong
Des MacHale, Cork
John MacNeill
C. Martschenko, Leicester
Michael Mudge, Wolverhampton
Stig Olsson, Sweden
A. R. Pargeter, Devon
Valerie de la Praudière, Winchester
E. S. Ratley, Essex
Martin J. Savier (?), Aberystwyth
Keith Selkirk, Nottingham
Alan Slomson, Leeds
Ian Stewart, Warwick
Tony Sudbery, York
Frank Tapson, Devon
G. Thompson, Peterborough
George Tyson, London
John Webb, Cape Town, South Africa
D. T. Whiteside, Cambridge
William Wynne Willson, Birmingham